Low Radar Cross Section
HIS-Based Phased Array

Low Radar Cross Section HIS-Based Phased Array

Radiation and Scattering Analysis

Hema Singh
Avinash Singh

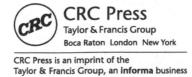

CRC Press
Taylor & Francis Group
Boca Raton London New York

CRC Press is an imprint of the
Taylor & Francis Group, an **informa** business

First edition published 2021
by CRC Press
6000 Broken Sound Parkway NW, Suite 300, Boca Raton, FL 33487-2742

and by CRC Press
2 Park Square, Milton Park, Abingdon, Oxon, OX14 4RN

© 2021 Taylor & Francis Group, LLC

CRC Press is an imprint of Taylor & Francis Group, LLC

ISBN: 978-0-367-51390-0 (hbk)
ISBN: 978-0-367-51394-8 (pbk)
ISBN: 978-1-003-05363-7 (ebk)

Typeset in Palatino
by Lumina Datamatics Limited

To

My beloved parents

Contents

List of Figures

List of Tables

Preface

Stealth technology relates to the reduction of infrared, optical and other electromagnetic signatures of combat aerospace platforms, warships, and vehicles in hostile environments. The reduction of radar cross section (RCS) essentially reduces the target detectability and hence, enhances its survivability. In aerospace platforms, the major contributor to RCS is the radiating antennas mounted over the platform. In other words, using passive methods of reducing RCS by shaping and using radar absorbing materials or structures would have no significance if the antennas/arrays mounted on the platform scatter energy of several orders of magnitude. Thus, it is necessary to focus on reducing the radar signatures from these antennas/arrays without any degradation in their radiation characteristics. Antennas such as dipoles, microstrip patches, etc., need a ground plane, which acts as a reflector to enhance the antenna gain/directivity. However, this metallic ground plane is one of the significant contributors of scattering from the antenna structure. This problem can be handled by incorporating high impedance surface (HIS) structures in the ground plane. Another challenge is to determine the antenna RCS when the antenna is radiating, referred to as radiation mode RCS. This book explains the electromagnetic (EM) design and performance analysis of such low profile microstrip patch arrays with HIS-based ground plane. Both radiation characteristics (gain, return loss, voltage standing wave ratio, VSWR) and radar cross section results are presented for different configurations of the patch arrays. Estimation of structural RCS and radiation mode RCS of the patch array of any arbitrary configuration have been explained. A systematic approach for design and realization of low RCS phased arrays along with numerous illustrations has been presented for the readers. This book would be of interest to beginners in the research area of antenna/array design toward low observable technology.

Acknowledgements

We would like to thank Mr. Jitendra J. Jadhav, Director, CSIR-National Aerospace Laboratories, Bangalore, for the permission to write this SpringerBrief.

We would also like to acknowledge valuable support from our colleagues at the Centre for Electromagnetics, Dr. R.U. Nair, Dr. Shiv Narayan, Mrs. Vineetha Joy, Mr. K.S. Venu, Mr. Ajith Nair, Ms. Deepa K. Sasidharan, Ms. Adrija Chowdhury, and Ms. Maumita Dutta during the course of writing this book.

Hema Singh would like to thank her daughter Ishita Singh for her constant cooperation and encouragement during the preparation of the book.

Avinash Singh would like to express his thanks to his parents Mr. Alok Singh and Mrs. Neelam Singh for their constant understanding and support during the writing of this book. He would also like to acknowledge his sister Ankita Singh for the sheer excitement and enthusiasm that she exuded toward this effort.

Authors

Dr. Hema Singh is working as senior principal scientist in Centre for Electromagnetics, National Aerospace Laboratories (CSIR-NAL), Bangalore, India. She received her PhD in electronics engineering from IIT-BHU, Varanasi India, in February 2000. For the period 1999–2001, she was a lecturer in physics at P.G. College, Kashipur, Uttaranchal, India. She was a lecturer in EEE of Birla Institute of Technology & Science (BITS), Pilani, Rajasthan, India, for the period 2001–2004. She joined CSIR-NAL as a scientist in January 2005. Her focus area of research is indigenous development of stealth technology, one of the critical challenges in the country's self-reliance in strategic sector. Moreover, phased arrays conformal to the platform are the present era requirement in view of stealth applications along with support to airborne/ground MTI radar and SAR for ISR mission applications. These advanced RF/IR/Optical sensors coupled with high-speed signal processing modules and algorithms provide the information beyond LOS. Dr. Singh is working toward design and development of low RCS low-profile phased array. She has been working on active RCS reduction, which relates to the real-time RCS reduction and control of aerospace structures. The antenna array mounted over a platform actively adapts the pattern in coherence with the overall structural RCS of the platform. Other contributions of Dr. Singh include EM analysis of propagation in an indoor environment for Boeing USA, HIS-based phased array design, RCS estimation of electrically large structures, and conformal arrays. She has authored or co-authored 12 books, 2 book chapters, 7 software copyrights, 360 scientific research papers, and technical reports.

Mr. Avinash Singh is a PhD scholar in the department of electronics and communication of Indian Institute of Technology (IIT), Roorkee, India. He has worked as a project assistant at the Centre for Electromagnetics (CEM) of CSIR-National Aerospace Laboratories, Bangalore for two and half years. He has obtained integrated BTech and MTech (Electronics and Communication) degrees from Amity University, Jaipur, India, in 2016. His research interests are phased arrays, corporate-fed planar/conformal patch array with conventional/HIS-based ground plane, active RCS reduction in phased arrays, estimation and control of RCS of phased arrays, EM design and analysis of multi-layered radar absorbing structures (RAS), and millimeter wave antenna arrays. He has co-authored 1 book chapter, 7 conference research papers, 18 technical documents, and 9 test reports.

About the Book

The design and development of low RCS phased array has been a challenging subject of interest in lieu of stealth technology. The underlying challenge is to reduce RCS of phased array without any degradation in the radiation characteristics. Over the years, various techniques have been developed for RCS reduction, such as using radar absorbing materials and structures (RAM/RAS), shaping, and active cancellation. The main challenge is to reduce both structural RCS and antenna mode RCS without affecting its directivity, gain, and efficiency. In a conventional low profile antenna design, a metallic ground plane acts as a reflector and also as a scatterer. Such metallic ground plane contributes to high scattering cross section of antenna. On the other hand, the frequency selective surface (FSS) elements, i.e., periodic metallic-dielectric structures, act as absorber in specific frequency band. When such periodic elements are included in the ground plane, they facilitate not only gain enhancement but also help in the reduction of antenna RCS. This book presents a comprehensive EM design and analysis of such low profile patch arrays with HIS-based ground plane. Readers would get to know how to determine radiation mode RCS of low profile antenna arrays with arbitrary configurations. Detailed descriptions of design, workflow of determining radiation and scattering behavior of antenna arrays have been supported with various schematics, tables and colored illustrations. This book would be of interest to antenna engineers working toward stealth technology.

1

Introduction

The design of stealth weapon systems such as combat aircraft or missiles requires low radar cross section (RCS) signatures. It involves the reduction of infrared, optical and other electromagnetic signatures of such aerospace platforms, warships and vehicles in hostile environments. One of the countermeasures to eliminate or reduce this threat is to minimize the RCS and thereby diminish the detectability of the target.

In aerospace vehicles, satellites, and similar systems, where size, weight, cost and performance are constraints, low profile antennas are used. Many government and commercial applications such as mobile, radio and wireless communications have similar specifications. However, in aerospace platforms, the major contributors to RCS are the radiating antennas mounted over such platforms. Thus, it is important to focus on reducing the radar signatures from these antennas/arrays without any disturbance in their radiation characteristics.

Over the years, various techniques have been developed for RCS reduction, such as using radar absorbing materials and structures (RAM/RAS), shaping and active cancellation. The main challenge is to reduce antenna RCS without affecting directivity, gain, efficiency and bandwidth of the antenna. The shaping techniques basically involve contouring the surface such that it redirects the scattered energy away from the source.

One way is to implement apertures and defected ground plane in the antenna design (Liu et al., 2014; Li et al., 2008). It is well known that the conformal patch arrays are preferred owing to their conformal geometry reducing the protruding structures from the platform. Furthermore, as compared to planar configurations, a conformal array scatters the incident EM wave in directions other than specular, thereby reducing the array RCS (Josefsson & Persson, 2006). RAM/RAS is comprised of coatings and composites that are capable of absorbing and attenuating incident waves within a particular frequency range. This helps in reducing backscattering. Frequency selective surfaces are periodic structures that selectively absorb or reflect a specific band of frequencies so as to provide out-of-band RCS reduction. The main advantage of the FSS structures is the planar arrangement, i.e., its low profile and the ability to conform to any surface while retaining its properties. These FSS structures can be incorporated in the design of microstrip patch antennas, which are also conformable to planar and non-planar surfaces, so as to reduce the RCS contribution of the patch antennas without deteriorating its radiation properties.

The electromagnetic bandgap (EBG) based radar absorbing materials (RAM) and frequency selective surface (FSS) based radar absorbing structure (RAS) have been incorporated in antenna design to achieve low RCS but with desired radiation characteristics. An ultrathin EM absorber has been proposed comprised of a single-layer treble square frequency selective surface (FSS) which resulted in broadband absorption bandwidth (Li et al., 2008). The active RCS reduction includes usage of active elements, plasma-based RCS reduction or adaptive probe cancellation. As far as passive RCS reduction is concerned, the structural RCS reduction is achieved in antenna/array(s). It is the radiation mode RCS, i.e., the antenna RCS while antenna is switched on, which needs to be estimated and hence controlled. This mode of antenna RCS is several times higher than the platform RCS over which the antenna is mounted. This brief focuses on control of both structural RCS and radiation mode RCS of patch arrays with the usage of HIS elements in the ground plane. A generic analytical formulation for the estimation of radiation mode RCS of phased array is described in subsequent subsections.

1.1 Need for HIS

A planar metal sheet is used as a ground plane in most antennas. The metal plate increases the gain of the antenna as it reflects most of the radiation in the opposite direction. Apart from this, metal has a unique property of supporting surface waves. In microwave frequencies these surface waves are just alternating currents (AC). In case of an infinitely large ground plane with smooth and plane sheet, surface waves do exist but do not contribute to any scattering. However, ground plane is a finite structure, and the surface waves will radiate into free space causing multipath interference if they encounter any discontinuities in shape.

RCS is often caused by the diffracted fields from the edges and corners of a structure, the surface waves travelling along the surface of the structure, waves traversing inside a cavity or partially closed structure and reflections from surfaces. Antenna placed on a platform is one of the major causes of this scattered energy. The scattered energy will be maximum when the antenna is radiating. There are instances where the RCS due to radiating antennas dominate over the RCS of platform.

One of the techniques to reduce the scattered energy from the antenna structure is to modify its ground plane. High impedance surface (HIS)-based ground plane can suppress these surface waves and thus reduce the scattered energy. HIS are the evolved versions of bumpy or corrugated surfaces. These textured surfaces alter the surface properties of the conductor (Sievenpiper et al., 1999). Periodic structures comprised of frequency selective surface (FSS) incorporated in the ground plane of antenna can facilitate RCS reduction. The period of these periodic structures should be smaller than the wavelength in order for the surface waves to encounter the textured surface. Inclusion of band-pass FSS in the ground plane results in a narrow out-of-band RCS reduction whereas band-reject FSS elements (e.g., octagonal loop resistive FSS) incorporated in the ground plane provide wideband RCS reduction.

One of the properties of the HIS cell is that it reflects the incident EM waves without any phase reversal and behaves as an artificial magnetic conductor (AMC). Based on this phenomena, researchers introduced a unique chessboard configuration. It consists of AMC and perfect electric conductor (PEC) cells. The principle of RCS reduction is the destructive interference between the reflected waves produced by the AMC and PEC cells. The AMC cells introduce a 0° phase shift while the PEC cell reflect the incident wave with 180° out-of-phase, resulting in a null in the specular direction. The HIS behaves as an AMC element only at the resonant frequency. Thus the nature of RCS reduction is narrowband. To improve narrow banding performance of the chessboard configuration, the PEC cells were replaced by another AMC element of same design configuration with different dimensions. Two AMC elements of different dimensions will have different resonant frequencies and the dimensions of the AMC elements are adjusted to achieve destructive interference for a broadband frequency range.

1.2 Different Configurations of HIS

This brief mainly discusses high impedance surfaces, the different arrangement of AMC cells in a high impedance surface and ways to implement a high impedance surface in an antenna/array system as a separate layer or integrate it with the ground plane itself. Figure 1.1 shows the various configurations of a HIS layer. A HIS layer comprised of AMC cells arranged uniformly over a substrate is shown in Figure 1.1a. On the other hand, Figure 1.1b shows a non-uniform arrangement of two AMC cells over a substrate. It can be observed from Figure 1.1c that the HIS layer consists of 2 × 2 elements of Jerusalem cross (JC) element as the AMC element and metallic square patch, arranged in a chessboard configuration. However, Figure 1.1d represents a modified ground plane. It is a hybrid configuration of the HIS layer arranged in a chessboard configuration and conventional metallic ground plane. The aim of the present work is to provide basic understanding about the integration of the high impedance surface layer in various conventional antenna/array systems and its role in radar cross section reduction.

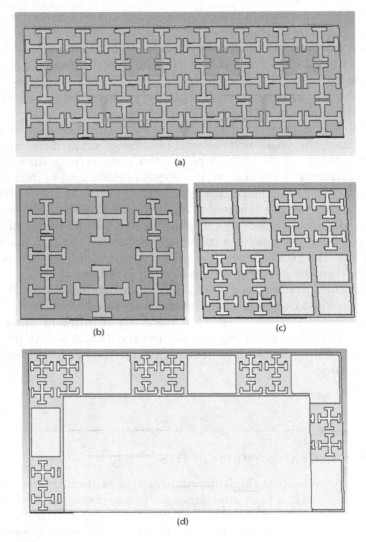

FIGURE 1.1
Design configuration of HIS layer: (a) Uniform HIS layer, (b) non-uniform HIS layer, (c) chess-board configuration of HIS layer, and (d) hybrid HIS layer.

While designing all the configurations, special care has been taken such that the inclusion of the HIS layer does not degrade the radiation characteristics of the antenna/array and at the same time RCS reduction (both structural RCS and radiation mode RCS) can be achieved. The structural RCS of antenna can be determined in a straightforward manner, because it relates to the scattering from the antenna array structure due to plane wave illumination. The main challenge is to calculate radiation mode RCS when the antenna array is excited.

1.3 Scattering Feature: Estimation of Radiation Mode RCS of Antenna Array

The in-band RCS of an antenna array is comprised of two scattering modes, the antenna (or the radiation mode) and the structural mode. The antenna mode depends on the radiation behavior of the antenna. It becomes negligible if the antenna is matched to its radiation impedance (Jenn, 1995). The structural mode arises from the currents induced on the antenna and the platform. In the case of a phased array, the structural RCS also signifies the case when the array is matched with the feed network.

The basic equation of antenna scattering (Jenn, 1995), which gives the total scattered field for a linearly polarized antenna, is expressed as

$$\vec{E}^S(Z_L) = \vec{E}^S(Z_a^*) + j\left[\frac{\eta_0}{4\lambda R_a}\vec{h}(\vec{h}\cdot\vec{E}^i)\frac{e^{-jk_oR}}{R}\right]\Gamma_0 \tag{1.1}$$

Here the antenna port is assumed to be terminated with load Z_L. The symbol * indicates complex conjugate. $\vec{E}^s$ is the scattered electric field, $\vec{E}^i$ is the incident electric field, $Z_a = R_a + jX_a$ is the radiation impedance with $R_a = R_r + R_d$.

The variables R_a is the antenna resistance, R_r is the radiation resistance, R_d is the ohmic resistance, Z_L is the load impedance, η_o is the free space impedance, h is the effective height of the antenna element, $k_o = 2\pi/\lambda$ is the free space wave number, R is the range between the target and the observation point and Γ_o is the reflection coefficient given by

$$\Gamma_o = \frac{Z_L - Z_a^*}{Z_L + Z_a^*} \tag{1.2}$$

In Equation (1.1), antenna RCS consists of two scattering modes, namely structural mode (first term) and antenna or radiation mode (second term). There are several situations when the radiation mode RCS becomes dominant, with negligible structural mode RCS. Mathematically, from Equation (1.2), if the load impedance (Z_L) is equal to the complex conjugate of antenna impedance (Z_a^*), the reflection coefficient Γ_o vanishes. This minimizes the reflections at the antenna terminals, hence the antenna RCS.

A typical example of a corporate feed (Figure 1.2) consists of couplers to either divide the power while in transmitting mode or couple/combine the power while in receiving mode. The feed network includes the phase shifters to achieve controlled electronic beam steering. Based on the respective impedances of each component, there are associated transmission/reflection coefficients. For example, the transmission coefficients t_r and t_p refer to that of the radiating element and the phase-shifters, respectively. Likewise, the reflection coefficients r_r, r_p and r_c correspond to the radiating element, the

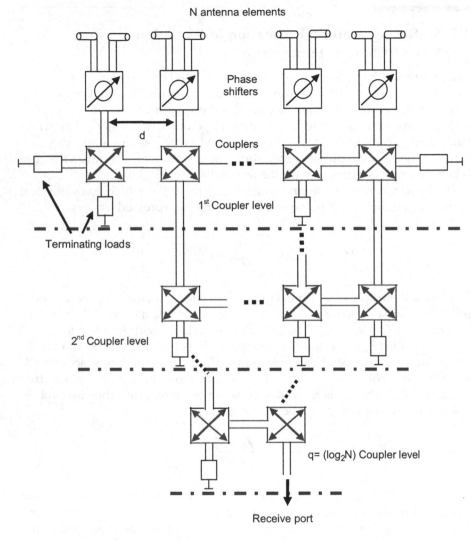

FIGURE 1.2
Typical corporate-fed linear phased array having radiating elements, phase shifters, couplers, and terminating loads.

phase-shifters and the couplers of the feed network, respectively. The number of antenna elements (N_x) in an array is related to the coupler level (q) in the feed network, i.e., $N_x = 2^q$. The level of couplers considered will have its effect on the overall array RCS. The coupler levels can be identified by the number of secondary lobes in the array RCS pattern (Jenn, 1995).

The total RCS of a phased array is comprised of the fields scattered due to the impedance mismatches at different junctions within the feed network.

The RCS is due to the scattering of incident signals that enter into the array aperture, travel within the feed network and then return back to the aperture after being reflected at various levels of the feed. The magnitude of these scattered fields is the measure of the reflection/transmission coefficients at their respective junctions expressed in the form of impedances.

The RCS for a linear array of N elements is expressed as

$$\sigma(\theta,\phi) = \frac{4\pi f^2 A_p^2}{c^2} \left| \sum_{i=1}^{N} \Gamma_i(\theta,\phi) e^{jk.\vec{d}_i} \right|^2 \left| F_{\text{normalized}}(\theta,\phi) \right|^2 \qquad (1.3)$$

while the RCS for a $N_x \times N_y$ planar array is given by

$$\sigma(\theta,\phi) = \frac{4\pi f^2 A_p^2}{c^2} \left| \sum_{i=1}^{N_x} \sum_{j=1}^{N_y} \Gamma_{ij}(\theta,\phi) e^{jk.\vec{d}_{ij}} \right|^2 \left| F_{\text{normalized}}(\theta,\phi) \right|^2 \qquad (1.4)$$

where σ is the RCS of phased array, $A_p = N_x N_y d_x d_y$ is the antenna aperture, $\vec{d}_i = \hat{x}(i-1)d$ is the position vector of antenna element in linear array, while for the planar array, it is $\vec{d}_{ij} = \hat{x}(i-1)d_x + \hat{y}(j-1)d_y$, Γ is the total reflection to the aperture, and $F_{\text{normalized}} = \cos\theta$ is the normalized element scattering pattern.

As discussed above, the sources of scattering in a given corporate-feed network are phase shifter inputs, r_p; aperture, r_r; input arms of first level couplers, r_c; sum and difference arm loads of the first level of couplers, r_{Σ_1} and r_{Δ_1}; and similarly higher level couplers, i.e., r_{Σ_i} and r_{Δ_i}, for $i = 2, 3\ldots$

The array factors depend on the array configuration, namely linear or planar. For a linear configuration having N_x elements along the x-axis, the sub-array factors associated with the components are given by

$$AF_r = r_r \left[\frac{\sin(N_x \alpha)}{N_x \sin\alpha} \right] \quad \text{for antenna elements} \qquad (1.5)$$

$$AF_p = r_p t_r^2 \left[\frac{\sin(N_x \alpha)}{N_x \sin\alpha} \right] \quad \text{for phase shifters} \qquad (1.6)$$

$$AF_c = \left(t_r t_p\right)^2 r_c \left[\frac{\sin(N_x \zeta_x)}{N_x \sin\zeta_x} \right] \quad \text{for side arms of 1st level couplers} \qquad (1.7)$$

$$AF_{\Sigma_1} = r_{\Sigma}\left(t_r t_p \cos\left(\frac{\zeta_x}{2}\right) \right)^2 \left(\frac{\sin\left(N_x \zeta_x\right)}{\frac{N_x}{2}\sin\left(2\zeta_x\right)} \right) \quad \text{for sum arms of 1st level couplers}$$

$$(1.8)$$

$$AF_{\Delta_1} = r_\Delta \left(t_r t_p \sin\left(\frac{\zeta_x}{2}\right) \right)^2 \left(\frac{\sin(N_x \zeta_x)}{\frac{N_x}{2}\sin(2\zeta_x)} \right)$$

(1.9)

for the difference arms of 1st level couplers

$$AF_{\Sigma_2} = r_\Sigma \left(t_r t_p \cos\left(\frac{\zeta_x}{2}\right)\cos\zeta_x \right)^2 \left(\frac{\sin(N_x \zeta_x)}{\frac{N_x}{4}\sin(4\zeta_x)} \right)$$

(1.10)

for the sum arms of 2nd level couplers

$$AF_{\Delta_2} = r_\Delta \left(t_r t_p \cos\left(\frac{\zeta_x}{2}\right)\sin\zeta_x \right)^2 \left(\frac{\sin(N_x \zeta_x)}{\frac{N_x}{4}\sin(4\zeta_x)} \right)$$

(1.11)

for the difference arms of 2nd level couplers

where $\alpha = -kd_x \sin\theta \cos\phi$, $\zeta_x = \alpha + \alpha_s$, r_Δ is the reflection coefficients of the difference output ports and r_Σ is the reflection coefficients of the sum output ports.

The scattering at various levels due to mismatches in the antenna feed network continues as the signal traverses deeper into the feed network. The vector sum of all the individual scattered fields that return to the aperture and re-radiate give the total RCS of the antenna array. The corporate feed network has been modeled with some assumptions. The elements are assumed to be non-ideal, i.e., the reflection/transmission coefficients r_r, r_p, r_c, t_r, t_p are non-zero entities. The scattering till second-level couplers are taken into account. Higher order reflections are considered negligible as the feed devices are assumed to have no reflections (i.e. $r \ll 1$) within the operating band. The effects of edges and mutual coupling are neglected. Lossless devices are assumed, i.e., $|r|^2 + |t|^2 = 1$. All couplers are modeled as Magic tees, in order to have uniform excitation.

The coherent sum of the scattered signals and hence the array RCS is represented by a non-coherent sum, assuming that only one term will be dominant for a given direction. Mathematically,

$$|E_1 + E_2 + ... + E_n|^2 \approx |E_1|^2 + |E_2|^2 + ... + |E_n|^2$$

(1.12)

The fraction of the signal entering the antenna array and returning to the aperture for re-radiation can be expressed as

$$\Gamma_{ij}(\theta,\phi) \approx r_r e^{j\Delta_{ij}} + t_r^2 r_p e^{j\Delta_{ij}} + (t_r t_p)^2 r_c e^{j2\chi_{ij}} e^{j\Delta_{ij}} + (t_r t_p t_c)^2 e^{j\chi_{ij}} \left[(E_1')_{ij} + (E_2')_{ij} + ... \right]$$

(1.13)

where $\Delta_{ij} = (i-1)\alpha + (j-1)\beta$ and $(E'_q)_{ij}$ is the reflected signal from qth level coupler to the element (i, j). The first and second terms of Equation (1.13) represent the contribution of radiating elements and phase shifters to the scattered field. The third and fourth terms collectively contribute to the RCS due to couplers. The respective expressions for the RCS of radiators, phase shifters, and couplers are analytically expressed as

$$\sigma_r(\theta,\phi) = F \sum_{i=1}^{N} r_{r_i} e^{j2(i-1)\alpha} \tag{1.14}$$

$$\sigma_p(\theta,\phi) = F \sum_{i=1}^{N} t_n^2 r_{p_i} e^{j2(i-1)\alpha} \tag{1.15}$$

$$\sigma_c(\theta,\phi) = F \sum_{i=1}^{N} t_n^2 t_{p_i}^2 r_{c_i} e^{j2(i-1)\zeta} \tag{1.16}$$

$$\sigma_{sd_1}(\theta,\phi) = F \sum_{i=1,3...}^{N-1} \left\{ \vec{E}_{i_1}^r(\theta,\phi) + \vec{E}_{(i+1)_1}^r(\theta,\phi) \right\} \tag{1.17}$$

$$\sigma_{sd_2}(\theta,\phi) = F \sum_{i=1,5...}^{N-3} \left\{ \begin{array}{l} \vec{E}_{i_2}^r(\theta,\phi) + \vec{E}_{(i+1)_2}^r(\theta,\phi) \\ \\ + \vec{E}_{(i+2)_2}^r(\theta,\phi) + \vec{E}_{(i+3)_2}^r(\theta,\phi) \end{array} \right\} \tag{1.18}$$

The total array RCS due to mismatches in the feed network considering scattering till second level of couplers is given by

$$\sigma(\theta,\phi) = \frac{4\pi f^2}{c^2} \left\{ \begin{array}{l} \left|\sigma_r(\theta,\phi)\right|^2 + \left|\sigma_p(\theta,\phi)\right|^2 + \left|\sigma_c(\theta,\phi)\right|^2 \\ \\ + \left|\sigma_{sd_1}(\theta,\phi)\right|^2 + \left|\sigma_{sd_2}(\theta,\phi)\right|^2 \end{array} \right\} \tag{1.19}$$

The detailed description of the analytical expressions for radiation mode RCS can be found in Singh & Jha (2015) and Singh et al. (2015a).

As an example, the broadside RCS of a linear 8-element microstrip patch array with a corporate feed network is shown in Figure 1.3. The length and width of the patch is $L = 8.22$ mm and $W = 11.22$ mm. The substrate is taken to be Arlon dielectric with a thickness $h = 1.588$ mm, relative permittivity $\varepsilon_r = 3.42$ and loss tangent $\tan\delta = 0.0035$. The interelement spacing between the patch is fixed at 0.5λ. It can be observed that the specular RCS is 25 dB at 8 GHz. The overall RCS pattern depends on several antenna parameters such as dimensions, inter-element spacing, frequency, impedances of the antenna array and its feed network. As already mentioned, the overall array RCS is the result of individual scattering contributions due to impedance mismatches throughout the array system.

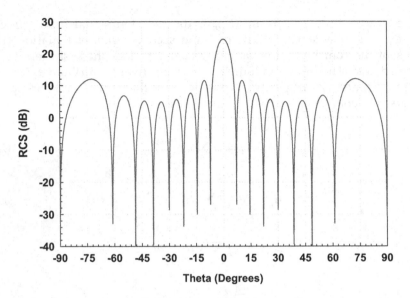

FIGURE 1.3
RCS of corporate-fed linear array with $N = 8$, $d = 0.5\lambda$, $f = 8$ GHz, $Z_O = 50\ \Omega$, $Z_L = 75\ \Omega$.

Organization of the book: This brief consists of six chapters. Chapter 1 gives a brief introduction about the importance of low RCS phased arrays in stealth technology. It describes how the conventional low profile patch arrays contribute to RCS by supporting surface waves and the need of HIS layer to suppress these surface waves. Chapter 2 discusses the EM design and analysis in terms of their radiation and scattering performance of the patch array with uniform and non-uniform layers of HIS in a planar patch array. Chapter 3 deals with the EM design and analysis of the chessboard configuration of the HIS layer along with hybrid HIS-based ground plane in a planar patch array as a three and two layered structure. Chapter 4 presents the EM design and performance analysis of a conformal patch array along with hybrid HIS-based ground plane toward broadband RCS reduction. Chapter 5 presents a low RCS proximity coupled patch array with a HIS layer for RCS reduction. Chapter 6 concludes the work done. For the sake of simplicity, the HIS layer used in all the design configurations consists of the same AMC element.

2

Uniform and Non-Uniform HIS Layer

Stealth technology involves the reduction of infrared, optical and other electromagnetic signatures of combat aerospace platforms, warships, and vehicles in hostile environments. The reduction of radar cross-section (RCS) essentially diminishes the detectability of the target. In aerospace platforms, the major contributor to RCS is the radiating antennas mounted over the platform. Thus it is important to focus on reducing the radar signatures from these antennas/arrays without any disturbance in their radiation characteristics.

Over the years, various techniques have been developed for RCS reduction, such as using radar absorbing materials and structures (RAM/RAS), shaping and active cancellation. The main challenge is to reduce antenna RCS without affecting directivity, gain, efficiency, and bandwidth of the antenna. The RCS reduction using shaping technique basically involves altering the shape of the structure/antenna under test (AUT) such that it redirects the scattered energy away from the target. One way is to implement apertures and defected ground plane in the antenna design (Liu et al., 2014; Li et al., 2008). The electromagnetic bandgap (EBG) based radar absorbing materials (RAM) and frequency selective surface (FSS) based radar absorbing structure (RAS) have been incorporated in antenna design to achieve low RCS but with desired radiation characteristics. Research (Li et al., 2012) has proposed an ultrathin EM absorber comprised of a single-layer treble square frequency selective surface (FSS) with broadband absorption bandwidth. A novel metallic EM structure as ground plane is used in antenna design (Sievenpiper et al., 1999). The proposed triangular lattice of hexagonal metallic plates, which is connected to a solid metal sheet by vertical conducting *vias*, poses a high surface impedance. This modified ground plane is referred to as HIS-based ground plane. In a conventional metallic ground plane, the antenna generates surface waves which further cause multipath interference and back scattering. However, in the presence of a HIS-based ground plane, the surface waves are suppressed, which results in the reduction in back scattering. Here HIS reflects the EM waves without any phase reversal and behaves as an artificial magnetic conductor (AMC).

Another thin HIS having a combination of AMC and perfect electric conductor (PEC) in a chessboard-like configuration shows remarkable performance in RCS reduction (Paquay et al., 2007; Iriarte et al., 2007). The advantage of using a chessboard configuration is the destructive interference between the reflected waves produced by the AMC and PEC cells. The AMC cells introduce a 0° phase shift while the PEC cells reflect the incident wave with 180° out-of-phase, resulting in a null in the specular direction. The disadvantage of the chessboard configuration of HIS is its narrowband behavior. For out-of-band frequencies, the AMC behaves as PEC with no destructive interference possible. This disadvantage was overcome by replacing the PEC cell with another AMC structure (Iriarte et al., 2013). It has been shown that such configuration can achieve RCS reduction over 40% bandwidth.

The work presented here is based on a similar concept; the RCS of a microstrip patch array is reduced by replacing the conventional ground plane with a Jerusalem cross- (JC-) based ground plane. The patch antenna is designed at a resonant frequency of 10 GHz. It has been taken care that due to low RCS, the radiation performance of patch antenna should not degrade. It is noted that the antenna RCS has two major contributions, namely the radiation mode RCS and the structural mode RCS.

2.1 Unit Cell Analysis

Band-stop FSS structures, if incorporated into the design of microstrip patch antenna as ground plane, could provide RCS reduction outside a specific frequency band. The resonant frequency and the bandwidth depend on the design of the FSS element and the properties of dielectric substrate (Iriarte et al., 2013). Jerusalem cross (JC), a band-stop FSS element, is considered in the present work.

Here, a substrate ($\varepsilon_r = 3.42$, $\tan \delta = 0.0035$) with thickness of 0.36 mm with JC-based FSS elements is used as ground plane. The schematic and dimensions of a JC element are shown in Figure 2.1, where L_{ind} and W_{ind} refer to length and width of the inductive element, and L_{cap} and W_{cap} refer to length and width of the capacitive element, respectively (Iriarte et al., 2013). A 2×2 array of FSS elements is designed, with inter-element spacing of 0.6 mm.

The reflection/transmission behavior of a JC-element for TE and TM polarizations are shown in Figure 2.2. The performance of patch antenna

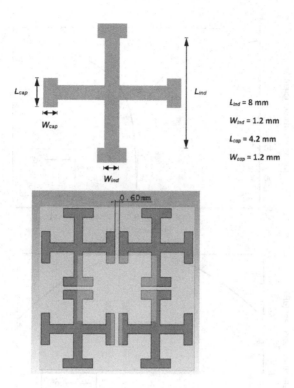

L_{ind} = 8 mm

W_{ind} = 1.2 mm

L_{cap} = 4.2 mm

W_{cap} = 1.2 mm

FIGURE 2.1
Schematic showing a JC-FSS array.

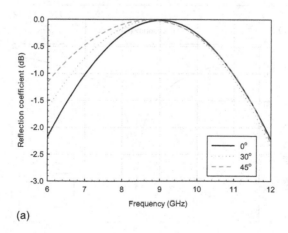

(a)

FIGURE 2.2
Reflection/transmission coefficients of a JC-FSS element: (a) TE-S_{11}. *(Continued)*

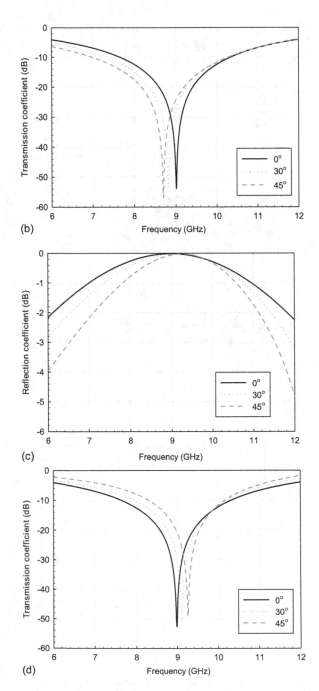

FIGURE 2.2 (Continued)
Reflection/transmission coefficients of a JC-FSS element: (b) TE-S_{21}, (c) TM-S_{11}, and (d) TM-S_{21}.

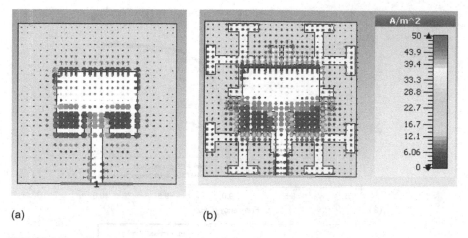

(a)   (b)

FIGURE 2.3
Surface current distribution over a single patch antenna: (a) PEC ground plane and (b) FSS-based ground plane.

with PEC ground plane is compared with the patch antenna with JC-based FSS ground plane. The surface current distribution over the patch antenna with and without the HIS layer is shown in Figure 2.3. The radiation and scattering characteristics of the single patch array is depicted in Figure 2.4.

Figure 2.4a shows the simulated return loss of the two patch antennas. It is apparent that the return loss is reduced in the FSS-based ground plane. Moreover, the resonant frequency is shifted from 9 to 9.008 GHz.

The patch antenna with PEC ground plane is −36.304 dB at 9 GHz with 2.73% bandwidth. The patch antenna with JC-based FSS ground plane shows return loss of −26.047 dB at 9.008 GHz with 3.54% bandwidth. This performance of patch antenna with JC-based FSS ground plane can be attributed to increase in the effective thickness of antenna. The far-field gain of both antennas is shown in Figure 2.4b. It is apparent that there is only slight variation in the antenna gain when the PEC ground plane is replaced with the FSS-based ground plane. Figure 2.4c presents the monostatic structural RCS of the single patch antenna with PEC ground plane and FSS-based ground plane at 9 GHz. It can be observed that the specular RCS is reduced by about 0.5 dB from −23.94 to −24.44 dBsm due to FSS-based ground plane.

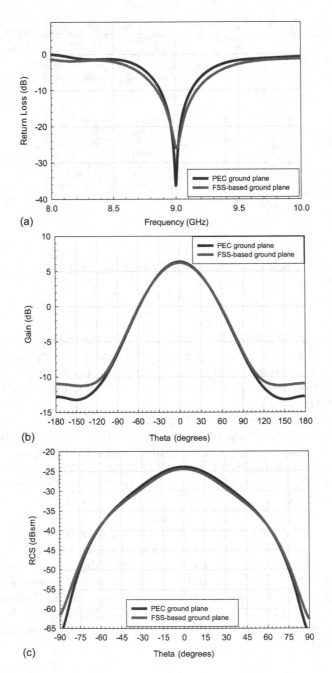

FIGURE 2.4
Radiation and scattering characteristics of single patch with JC HIS layer: (a) Return loss, (b) gain, and (c) structural RCS pattern.

2.2 Patch Array with Uniform HIS Layer

The design of single patch antenna for 9 GHz is extended to 4-element and 8-element linear patch arrays. While doing so, a feed network has to be incorporated to provide a common port. Here, a microstrip line-based corporate feed network is used to provide proper impedance matching.

Figure 2.5a shows a 4-element linear microstrip patch array with conventional PEC ground plane and corporate feed network. The patch dimensions are the same as that of the single patch described in the previous section. The center-to-center separation between adjacent patch elements is taken as 0.66λ. A two-level corporate feed network is designed, with the feed dimensions computed using

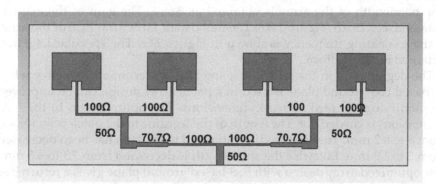

(a)

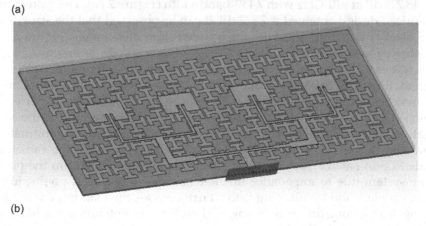

(b)

FIGURE 2.5
(a) Design of 4-element linear microstrip patch array with corporate feed network. (b) A 4-element linear patch array with uniform HIS-based ground plane.

$$Z_0 = \eta_0 \frac{1}{\sqrt{\varepsilon_r'}\left(1.393 + \frac{W_f}{h} + \frac{2}{3}\ln\left(\frac{W_f}{h} + 1.444\right)\right)} \qquad (2.1)$$

where η_0 is the impedance of free space, ε_r' is the effective dielectric constant, W_f is the width of microstrip feed line, and h is the thickness of the substrate.

As shown in Figure 2.5a, a single feed matched to 50 Ω is branched out to two 100 Ω feedlines, each of which is then connected to a 2-element patch array. The 100 Ω line is matched to 50 Ω line through a 70.7 Ω stub. The 50 Ω feed is again branched out to 100 Ω lines, which are finally connected to each patch element. Figure 2.5b shows the 4-element patch array with uniform HIS inserted as a middle layer above the ground plane.

The designed linear patch array with conventional ground plane resonates at 8.992 GHz with a return loss of −50.515 dB, as shown in Figure 2.6a. The bandwidth of the array is obtained as 3.69%. The gain of the array is obtained as 9.23 dB (Figure 2.6b). The monostatic structural RCS of the array at the resonating frequency is shown in Figure 2.6c. The specular lobe RCS obtained is −10.04 dBsm.

The degradation in the return loss and shift in resonant frequency when JC-based FSS ground plane is used in a patch array design can be improved by modifying the feed network dimensions. The optimization in the feed dimensions is carried out. The length of the feedline to the patch is increased from 8 to 8.2 mm. The 50 Ω feedline attached to the stub has been decreased from 2 to 1.9 mm. Likewise the stub length is decreased from 7.3 to 6.3 mm. The optimized array design with FSS-based ground plane gives a return loss of −48.721 dB at 9.01 GHz with 7.43% bandwidth (Figure 2.6a). The gain of the optimized design obtained is 7.823 dB. It can be observed that the structural RCS of array with optimized design does not show any change (Figure 2.6c).

Estimation of Radiation Mode RCS: Radiation mode or antenna mode RCS relates to the scattering of plane waves when the antenna feed point is excited. It mainly depends on the radiation characteristics and impedance mismatches within the antenna array system. If the antenna is perfectly matched (i.e., $Z_L = Za^*$, where, Z_L is the load impedance and Za^* is the conjugate of antenna impedance), the radiation mode RCS can be eliminated. However, this is not practically achievable. This is because the radiation mode RCS in patch array is a result of multiple reflections within the patch array system due to impedance mismatches at antenna feed points, junctions, couplers and terminating loads. Furthermore, the load impedance for conjugate matching differs with angle. Therefore, the antenna mode RCS has a significant contribution toward the total array RCS.

In order to determine radiation mode RCS, one need impedances at each junction of the phased array system. Here a novel method is presented to calculate the antenna impedance of each patch element in a multi-resonant antenna array structure in order to compute the radiation mode RCS.

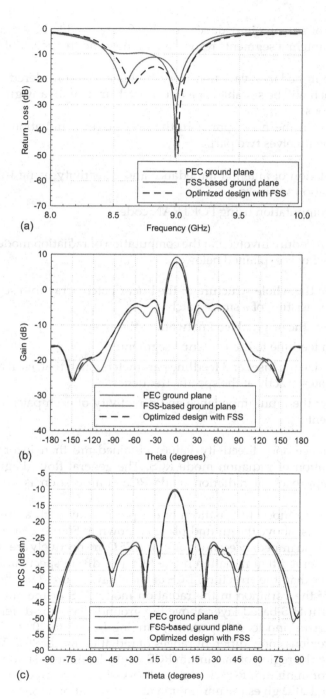

FIGURE 2.6
Radiation and scattering characteristics of patch array with and without JC-based HIS layer:
(a) Return loss, (b) gain, and (c) structural RCS pattern.

The values of antenna impedance and directivity need to be extracted from individual antenna segments resonating at the same frequency of the whole patch array.

Since the impedance variation in each element is considered separately, this approach will be suitable for a linear patch array with any ground plane configurations.

According to the new approach, the procedure of radiation mode RCS computation involves two parts:

1. Calculation of antenna impedance and directivity using full wave simulation software.
2. RCS computation using FORTRAN code:

The basic procedure involved in the computation of radiation mode. The RCS of a patch array is explained below:

- Divide the whole structure of the linear patch array into segments, each consisting of a single patch.
- Remove the original feed network of the array.
- Add a feedline to each antenna segment.
- Optimize the inset and feedline parameters, such that, each antenna resonates exactly at the specific frequency.
- Extract the input impedance and directivity of each patch antenna segment.

The impedance and directivity values obtained are then incorporated in the formulation of radiation mode RCS. The general flow diagram of the proposed approach of radiation mode RCS computation is illustrated in Figure 2.7.

The radiation mode RCS pattern of the patch array with a uniform HIS-based layer just above the ground plane is shown in Figure 2.8a. The characteristic and load impedance values of 50 and 75 Ω, respectively are considered. It can be observed that radiation mode RCS (9.0 dB) in specular direction is much higher than the structural RCS of the array. Figure 2.8b shows a contour plot of the variation in the radiation mode RCS of the antenna array with uniform HIS-based layer above the ground plane for different values of characteristic and load impedance, ranging from 5 to 200 Ω. The different shade regions show different ranges of radiation mode RCS. The darkest shade being the region of maximum RCS while the lightest shade represents the region of minimum RCS. It can be observed that an impedance range of around 110–120 Ω gives the minimum value of radiation mode RCS.

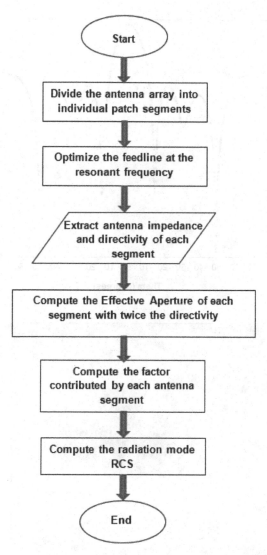

FIGURE 2.7
Flowchart for computation of radiation mode RCS of a patch array.

While analyzing the results, it can be noted that the design configuration of a patch array with uniform HIS layer is unable to provide RCS reduction at resonant frequency. In the next section, a different approach has been taken to analyze the scattering performance patch array with HIS layer when arranged in a non-uniform configuration/pattern.

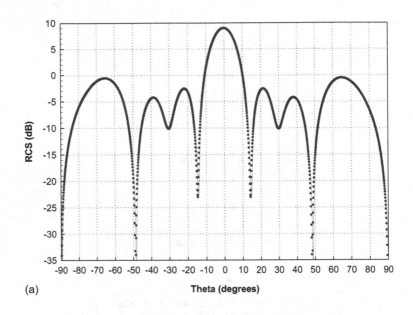

(a)

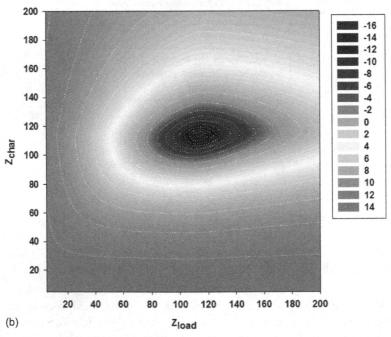

(b)

FIGURE 2.8
(a) Radiation mode RCS pattern of 4-element patch array with uniform HIS layer. (b) Contour plot for radiation mode RCS.

2.3 Patch Array with Non-Uniform HIS Layer

In this section, two JC elements of different dimensions are taken into consideration. The EM design of each JC unit cell, each resonating at a unique frequency, is discussed. A unit cell essentially is comprised of a single FSS element backed by a dielectric substrate. The substrate for both the designs is taken the same, so as to analyze the scattering behavior for a combination of the two FSS elements. The effect of varying the thickness of dielectric and dimensions of the JC element, on transmission and reflection coefficients, and reflection phase are also discussed. Figure 2.9 shows the schematic of two JC-FSS unit cells, with dimensions mentioned in Table 2.1.

A 0.2 mm thick FR4 material, with relative permittivity (ε_r) of 4.3 and loss tangent (tan δ) of 0.025, is taken as the dielectric substrate. Figure 2.10a and b show the reflection (S_{11}) and transmission (S_{12}) coefficients of the two JC-based FSS elements.

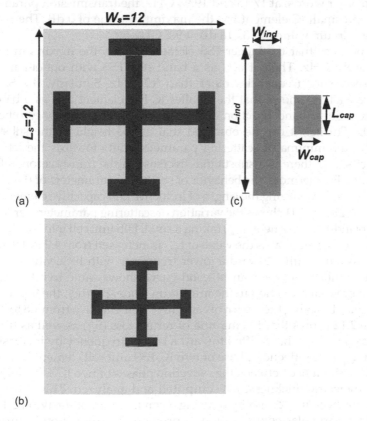

(a)

(b)

(c)

FIGURE 2.9
Schematic of JC-FSS unit cells on dielectric substrate: (a) Big JC-FSS unit cell, (b) small JC-FSS unit cell, and (c) inductive and capacitive elements of a JC-FSS.

TABLE 2.1

Dimensions of Jerusalem Cross (JC) Elements in FSS
Unit Cell

Dimension	Small JC	Big JC
L_{ind}	5.4 mm	8 mm
W_{ind}	0.8 mm	1.2 mm
L_{cap}	2.4 mm	3.5 mm
W_{cap}	0.8 mm	1.2 mm

It can be observed that the bigger FSS element shows minimum reflection at 17.13 GHz with a return loss of −67.07 dB, while the smaller element shows minimum reflection at 19.94 GHz with a return loss of −46.55 dB, provided the height of the substrate over which both the FSS elements have been designed is kept 0.2 mm. As observed in Figure 2.10a (ii), the transmission parameter is complementary to the reflection parameter for a given FSS element. In other words, at 17.13 and 19.94 GHz, the transmission parameter of the big and small JC element has the maximum value of 0 dB. The S_{12} curve shows minimum value of −35.18 dB at 9.2 GHz.

This indicates that the bigger FSS element reflects the maximum incident energy at 9.2 GHz. Thus, it acts as a band-stop FSS with out-of-band rejection observed for frequencies other than 9.2 GHz. Similarly, the S_{12} curve of Figure 2.10b (ii) shows that the smaller JC-FSS element acts as a band-stop FSS, where out-of-band rejection can be observed for a frequency other than 13.6 GHz. Further, it can be observed that as the height of the substrate is increased, the minima of scattering parameter shifts towards the left (lower frequencies) and there is a substantial decrease in the transmission/reflection parameters. Furthermore, the behavior of scattering parameters of the HIS unit cell is analyzed by varying the length of inductive and capacitive elements of the JC element. Figure 2.11 shows the variation of scattering parameters for different lengths of inductive element (L_{ind}) taking a small HIS unit cell into consideration.

It can be observed that as the value of L_{ind} is increased from 4.2 to 4.9 mm, the return loss curve shifts toward a lower frequency with little variation in the magnitude. Likewise, the transmission curve shows same trend except that there is a gradual reduction in the minimum value. Further, the S-parameters are computed varying the length of capacitive element (L_{cap}) from 1.8 to 2.2 mm.

Figure 2.12 shows that the minima of return loss (S_{11}) as well as transmission parameter (S_{12}) has shifted toward a higher frequency by increasing the value of L_{cap}. The reflection phase of two JC-FSS unit cell elements is also analyzed. The difference between the reflection phases of two JC-FSS element for various dielectric thicknesses is computed and analyzed. This is calculated in order to check the feasibility of using a combination of the two JC-FSS elements in a particular configuration to provide RCS reduction in microstrip patch array. This can be attributed to the fact that if the reflections come from

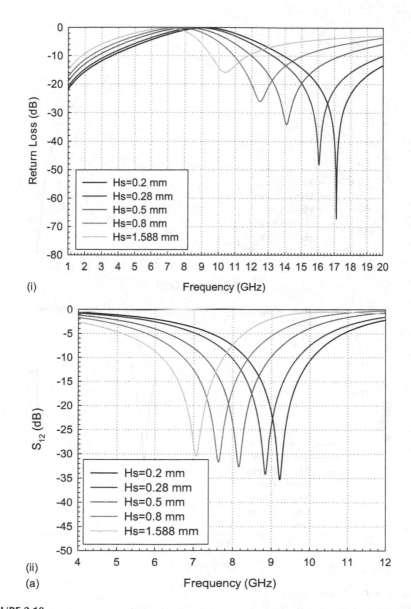

FIGURE 2.10
(a) Scattering parameters of big JC-FSS unit cell. (i) Return loss, S_{11} (ii) Transmission parameter, S_{12}. *(Continued)*

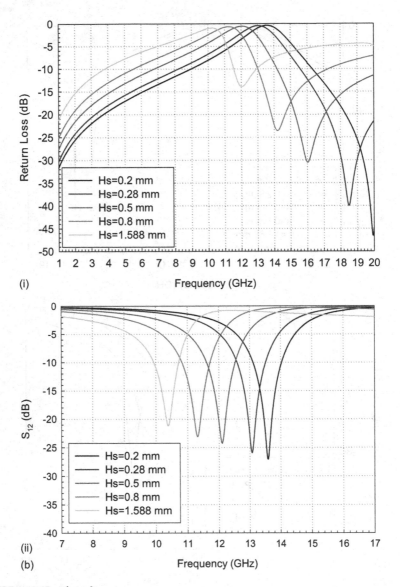

FIGURE 2.10 (Continued)
(b) Scattering parameters of small JC-FSS unit cell. (i) Return loss, S_{11} (ii) Transmission parameter, S_{12}.

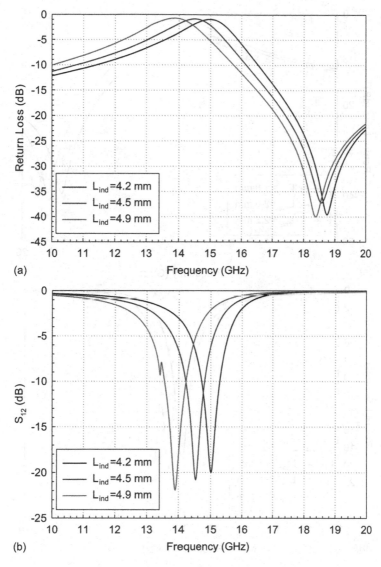

FIGURE 2.11
Scattering parameters of big JC-FSS unit cell for different values of L_{ind}: (a) Return loss, S_{11} and (b) transmission parameter, S_{12}.

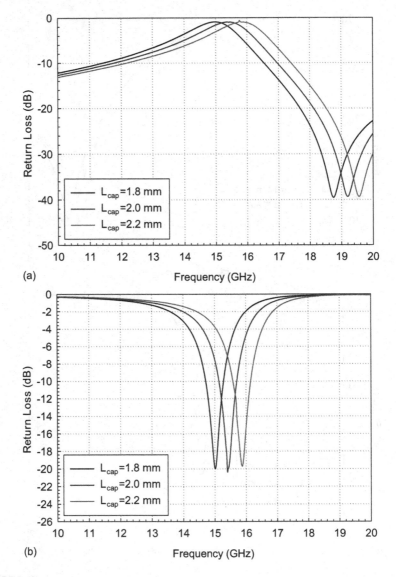

FIGURE 2.12
Scattering parameters of big JC-FSS unit cell for different values of L_{cap}: (a) Return loss, S_{11} and (b) transmission parameter, S_{12}.

both FSS unit cells, backscattering or structural RCS (in specular direction) for that frequency band will be nullified.

The analysis was done for different values of substrate thicknesses. Reflection phase (Figure 2.13a) and phase difference are computed for the two JC-FSS unit cells for dielectric thickness of 0.2 mm as well. From Figure 2.13b, it is clear that −180° phase difference is attained from approximately 17 to 20 GHz, which proves that the design is suitable for reducing backscattering

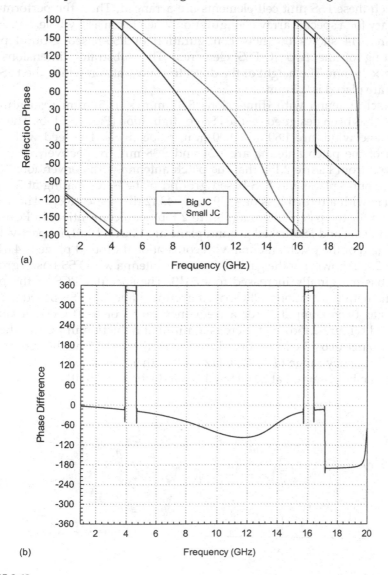

(a)

(b)

FIGURE 2.13

(a) Reflection phase of JC-FSS unit cells and (b) reflection phase difference between big and small JC-FSS unit cells.

over the specified frequency range. Among all the values, it was observed that 0.2 mm thick FR4 substrate is capable of providing RCS reduction over a wide frequency band when compared to other substrate thicknesses. The inference has been made with respect to the scattering behavior of two JC-FSS elements computed individually. The results obtained above are not sufficient to ensure the fact that total RCS reduction will be achieved for a frequency range where out-of-band phase reflection is obtained. It also depends on the configuration in which these FSS unit cell elements are arranged. Thus, the performance may vary for a specific array configuration when taken as a whole.

Figure 2.14 shows the new configuration of FSS-based ground plane consisting of the two JC-FSS elements. The substrate dimensions are 26 mm × 36 mm. The height of the dielectric substrate on which the FSS elements are placed is 0.2 mm.

A patch antenna with dimensions 8.37 mm × 11.51 mm resonating at 8.0 GHz has been designed on the FSS configuration (Figure 2.15b). The inset length and width are 1.80 and 1.00 mm, respectively. The feed width and length of the patch antenna are 1.24 and 9.38 mm, respectively. It can be observed from Figure 2.15a that the patch antenna with conventional PEC-based ground plane and FSS-based ground plane is resonating at 7.98 and 7.93 GHz with a return loss of −37.9 and −43.12 dB, respectively. The VSWR obtained for the patch antenna (Figure 2.15b) with conventional PEC-based ground plane and FSS-based ground plane are 1.02 and 1.01, respectively.

The gain of the patch antenna with conventional ground plane is 4.64 dB. Figure 2.15c shows that the gain of the patch antenna with FSS-based ground plane has marginally increased to 4.7 dB. The specular RCS of the patch antenna with conventional PEC-based ground plane and FSS-based ground plane has been computed for a frequency range of 6–20 GHz. It can be observed (Figure 2.15d) that there is a minimal reduction in RCS in the frequency range of 6–11 GHz. A broadband reduction in RCS has been observed in the frequency range of 14–20 GHz and further. The maximum reduction in RCS can be observed at 20 GHz with a reduction of 2.3 dB.

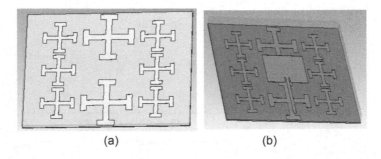

(a)   (b)

FIGURE 2.14
(a) Schematic of the FSS array configuration and (b) single patch antenna on FSS-based ground plane.

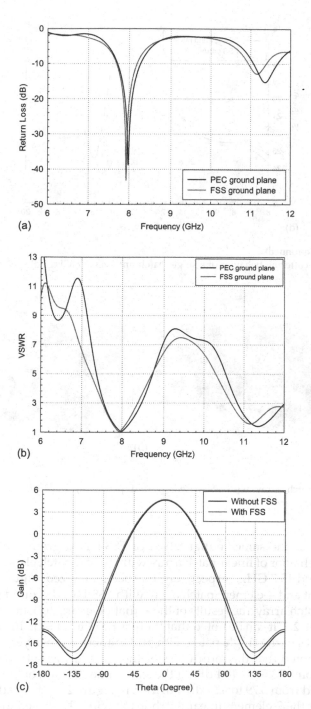

FIGURE 2.15
Radiation and scattering behavior of a single patch antenna at 8 GHz: (a) Return loss, (b) VSWR,
(c) gain. *(Continued)*

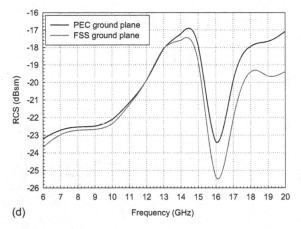

(d)

FIGURE 2.15 (Continued)
Radiation and scattering behavior of a single patch antenna at 8 GHz: (d) specular structural RCS.

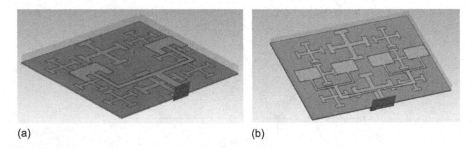

(a)   (b)

FIGURE 2.16
Schematic of patch array with FSS-based ground plane: (a) 2-element patch array and (b) 4-element patch array.

The effect of the same configuration of FSS-based ground plane on the scattering behavior of linear patch array with 2 and 4 patch elements resonating at 13.5 and 15.7 GHz, respectively, has been analyzed. Figure 2.16 shows the 2-element and 4-element patch array with FSS-based ground plane. For a 2-element patch array, the results of the radiation characteristics are summarized in Table 2.2. It can be noted that there is no degradation in the radiation characteristics of the single patch with and without a HIS layer.

It can be noted that the return loss of the patch array with HIS layer has degraded from −33.5 to −18.1 dB. At the same time the gain of the patch array has increased from 5.79 to 6.57 dB. Further, in Figure 2.17, the scattering characteristics of the 2-element linear patch array with PEC-based ground plane are compared with that with FSS-based ground plane. In this case, reduction in RCS can be observed from 7 to 20 GHz. The maximum RCS reduction

TABLE 2.2
Radiation Characteristics of a 2-Element Microstrip Patch Array at 13.5 GHz

Array Performance	PEC Ground Plane	FSS-Based Ground Plane
Return loss	−33.5 dB at 13.53 GHz	−18.1 dB at 13.57 GHz
VSWR	1.04	1.28
Gain	5.79 dB	6.57 dB

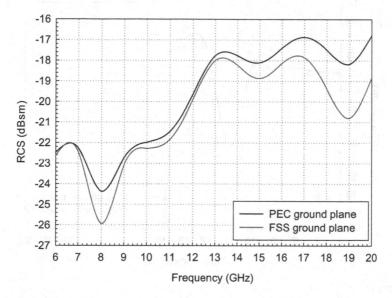

FIGURE 2.17
Specular structural RCS of 2-element linear patch array resonating at 13.53 GHz.

achieved is 2.7 dB at 19 GHz. It can be inferred from the above results that this configuration of FSS-based ground plane can be further improved with optimized design parameters.

Similar analysis is carried out for a 4-element linear patch array resonating at 15.7 GHz. The radiation characteristics are shown in Table 2.3. In this case, the return loss of a 4-element linear patch array with a conventional PEC ground plane obtained is −44.39 dB at 15.72 GHz and with FSS-based ground plane, the return loss achieved is −27.61 dB at 15.42 GHz. The VSWR has increased marginally from 1.01 to 1.08. The gain of a 4-element linear patch array with PEC based ground plane obtained is 7.69 dB and has increased to 8.03 dB with FSS-based ground plane. The scattering characteristics of the 4-element linear patch array with PEC-based ground plane are compared to with that of FSS-based ground plane in Figure 2.18. The maximum RCS reduction achieved is 8 dB at 11 GHz.

TABLE 2.3
Radiation Characteristics of a 4-Element Microstrip Patch Array at 15.7 GHz

Array Performance	PEC Ground Plane	FSS-Based Ground Plane
Return loss	−44.39 dB at 15.72 GHz	−27.61 dB at 15.42 GHz
VSWR	1.01	1.08
Gain	7.69 dB	8.03 dB

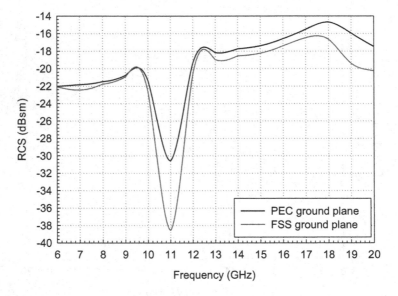

FIGURE 2.18
Specular structural RCS of 4-element linear patch array resonating at 15.7 GHz.

2.4 Summary

The low observable technology strongly demands the low RCS antenna/ array mounted on the aerospace platforms. The total RCS of the platform has a major contribution from the antennas mounted over it. Further it is expected that the low RCS antenna/array should have good radiation characteristics. In this chapter, the ground plane of the patch array is modified to achieve reduction in array RCS without any degradation in the radiation performance. The conventional PEC ground plane is replaced with uniform and non-uniform configuration of Jerusalem cross (JC) elements in the HIS layer. For uniform FSS configuration, although there was no degradation in the radiation characteristics, no structural RCS reduction was observed at the resonating frequency. Further, two JC-based FSS unit cells are designed, on FR4 substrate, at two resonating frequencies such that their corresponding reflection phase cancels out each other for a certain frequency range. The bandwidth and the magnitude of reduction in RCS depend on the dielectric thickness and the arrangement of FSS elements in the array. The scattering behavior of the microstrip patch array is analyzed by using these JC-FSS unit cells in non-uniform configuration as ground plane. It is observed that the frequency band which provides maximum RCS reduction varies with each antenna even for the same FSS configuration. For the selected FSS configuration, maximum reduction in specular structural RCS has been observed in the 4-element patch array resonating at 15.7 GHz, while relatively wideband RCS reduction has been observed in the 2-element patch array resonating at 13.5 GHz.

3

Hybrid HIS-Based Ground Plane

Antenna radar cross section (RCS) is a significant parameter to be considered in a low observable platform. The antenna RCS is essentially a combination of both radiation mode and structural mode RCS. The total RCS of an antenna is given by (Hansen, 1989)

$$\sigma = \left| \sqrt{\sigma_{struct}} - (1-\Gamma)\sqrt{\sigma_{rad}}\, e^{j\varphi} \right|^2 \tag{3.1}$$

where σ_{struct} and σ_{rad} represent the structural mode and radiation mode RCS, respectively, Γ is the reflection coefficient of the antenna, and φ is the relative phase between σ_{struct} and σ_{rad}. The structural mode RCS is the back scattering only by the antenna structure when it is not in operation while the RCS contributed by the antenna when it is switched on is the radiation mode RCS. The radiation mode RCS of the antenna/array is null when the antenna is perfectly matched to the load. The present work is aimed at reducing the structural RCS of a microstrip patch array.

As previously mentioned, a combination of artificial magnetic conductor (AMC) and PEC cells in a chessboard-like configuration exhibit RCS reduction. The PEC cells reflect the incident waves with 180° phase shift, while AMC cells introduce a 0° phase shift. The combination of both PEC and AMC cells leads to phase cancellation, and hence the RCS reduction. The main drawback of this configuration is that it provides narrowband RCS reduction. However, to overcome this narrowband behavior the PEC cells are replaced with the same AMC cell but with different dimensions. Thus, the two AMC cells with different dimensions resonated at two different frequencies, and hence provide reflection phase cancellation to the incident wave over the broadband frequency range.

In this chapter, broadband RCS reduction is achieved using only one AMC element. Moreover, the design is such that there is no degradation in the antenna radiation characteristics. Two different design configurations are presented in this chapter. The three-layered structure has a HIS layer in the middle along with the hybrid ground plane. As the three-layered design is complex from fabrication point of view, later a two-layered design has been proposed comprising of only HIS-based hybrid ground plane.

3.1 Chessboard Configuration

Band-stop FSS structures, designed at a particular operating frequency, act as a perfect electric conductor (PEC) and hence it results in maximum reflection. At other frequencies, they allow incident waves to partially pass through them. Therefore, they can provide out-of-band RCS reduction (Munk, 2000). The FSS elements and metallic square patches can be combined in different configurations to form HIS-based layer in order to achieve wideband RCS reduction. These HIS layers can be incorporated in patch antenna/array as substrate or ground plane to reduce its structural RCS.

In this section, the design and performance analysis of an HIS unit cell composed of Jerusalem cross (JC) elements and metallic square patches is discussed. Furthermore, the radiation and scattering characteristics of a single patch antenna with the designed HIS-based ground plane is presented. The characteristics of the HIS structure are studied by defining a unit cell. Figure 3.1 shows the schematic of the HIS unit cell and JC element. It is a combination of JC and square patch elements arranged in a chessboard configuration, backed by a dielectric substrate of 30×30 mm dimension. The JC elements are arranged in a 2×2 configuration. A 0.28 mm thick FR4 material ($\varepsilon_r = 4.3$, tan $\delta = 0.025$) is chosen as the dielectric substrate. Copper ($\sigma = 5.8 \times 10^7$) is taken as the material for all metallic parts. The length of the square patch is 13.9 mm.

The length and width of the inductive element are 4.8 and 0.8 mm, while for the capacitive element the length and width are 2.8 and 0.8 mm, respectively. The spacing between two JC elements, as well as between a square patch and a JC array is 1.1 mm. Figure 3.2 shows the variation of reflection phase of an HIS unit cell with frequency. It can be observed that a transition

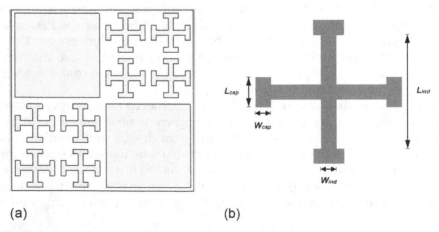

(a)   (b)

FIGURE 3.1
(a) HIS unit cell and (b) schematic of JC element.

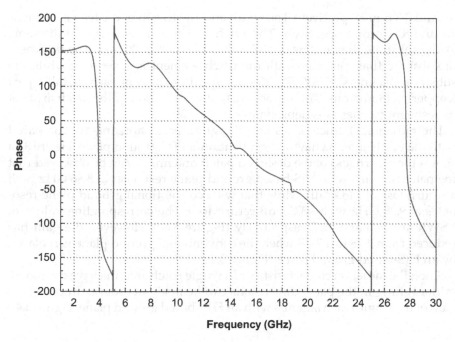

FIGURE 3.2
Reflection phase of HIS unit cell.

from $+180° \pm 30°$ to $-180° \pm 30°$ has given rise to zero points of reflection near 4, 15 and 28 GHz. At these frequencies, it is expected that the HIS layer will provide maximum RCS reduction.

A single patch antenna on an FR4-based substrate resonating at 9 GHz is shown in Figure 3.3a. The patch dimensions for the specified resonant frequency are 7.34×10.24 mm. The optimized length and width of the inset are

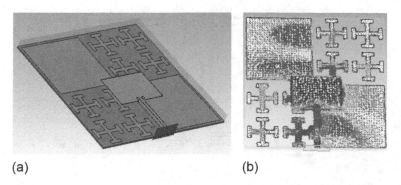

(a)   (b)

FIGURE 3.3
(a) A single microstrip patch antenna with HIS-based ground plane with resonating frequency of 9 GHz and (b) surface current distribution over the patch antenna and HIS layer.

0.9 and 0.5 mm respectively. The width and the length of the feed is taken as 0.7 and 8.33 mm, respectively. The antenna dimensions are taken the same as the substrate dimensions of HIS unit cell, i.e., 30 × 30 mm. The thickness of substrate that comes beneath the patch element is 1.588 mm, while the substrate thickness of HIS is 0.28 mm. The thickness of each metallic part (copper) is 18 microns. The surface current distribution of the patch array at its resonant frequency is depicted in Figure 3.3b.

The radiation characteristics of a designed patch antenna are compared with the patch antenna having a conventional PEC ground plane. Figure 3.4a shows the return loss of the designed patch antenna. It can be observed that the patch antenna with HIS-based ground plane resonates at 8.86 GHz with a return loss of −18.47 dB, while that with conventional ground plane reso-nates at 9.08 GHz with −37.77 dB return loss. The corresponding values of VSWR are 1.03 and 1.27, respectively (Figure 3.4b). However, the gain has reduced from 4.78 to 3.3 dB when the conventional ground plane is replaced by an HIS-based ground plane (Figure 3.4c).

Since, the radiation characteristics of a single patch antenna have degraded, minor tuning in the patch dimensions can be made in order to improve the gain of the single patch antenna with an HIS-based ground plane. Figure 3.4d

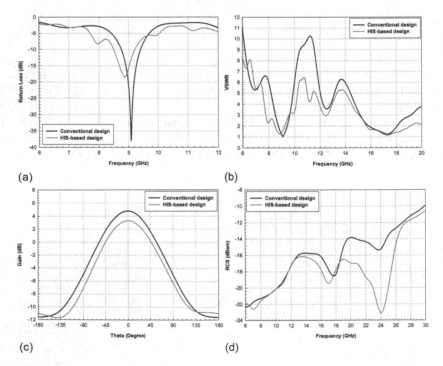

(a)   (b)

(c)   (d)

FIGURE 3.4
Radiation and scattering characteristics of a single patch antenna: (a) Return loss, (b) VSWR, (c) gain, and (d) specular structural RCS.

shows the structural RCS of a single patch antenna with an HIS-based ground plane. It is evident that the RCS reduction is achieved in the frequency range of 13–17 GHz and 18.5–30 GHz. The maximum RCS reduction (7.9 dBsm) is obtained at 24 GHz. There is no reduction in the RCS around 10 GHz. This is because the HIS acts as a perfect reflector at the resonant frequency of the antenna. It is found that the antenna resonates at 18 GHz as well. This can be the reason that the RCS of the patch array with HIS layer is comparable to the RCS of the reference patch at 18 GHz (Figure 3.4d). However, since the radiation characteristics have degraded, which is undesirable, a different design configuration of HIS layer is considered.

3.2 Combination of HIS Layer and Hybrid Ground Plane: Three-Layered Structure

This configuration is comprised of a microstrip patch array with a ground plane consisting of two metallic layers separated by a dielectric substrate. The structure of the array can be described in terms of three layers as depicted in Figure 3.5. The top layer comprises the metallic patch array on a

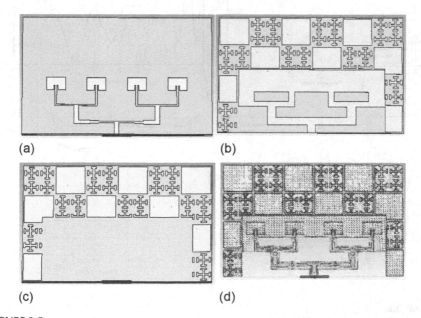

(a) (b) (c) (d)

FIGURE 3.5
Three-layered design: A 4-element linear patch array with non-uniform HIS layer and modified ground plane at 10.5 GHz (Substrate: **60 × 90 mm**): (a) Top layer, (b) middle layer, (c) bottom layer, and (d) surface current distribution at resonant frequency.

dielectric (FR4) substrate of thickness 1.588 mm. A 4-element microstrip patch array resonating at 10.5 GHz and with substrate dimensions 60 × 90 mm is considered. The middle layer is a hybrid HIS layer, consisting of a chessboard configuration of JC-square patch elements and a metallic layer on a FR4 dielectric material of 0.28 mm thickness (Figure 3.5b). The dimensions of the JC element are the same as in Section 3.1.

The JC elements and square patches are made up of copper of thickness 18 microns. It can be noted from Figure 3.5c that the ground plane consists of only the HIS layer. The portions that come directly below the patch array are removed in the middle and ground plane. The metallic plate in the middle layer has an outline structure of the patch array.

The radiation and scattering characteristics of the designed patch array are shown in Figure 3.6. The resonant frequency has shifted from 10.48 to 10.33 GHz. The return loss of a conventional patch array is −26.68 dB while that of the patch array with hybrid HIS layer and modified ground plane is −23.34 dB.

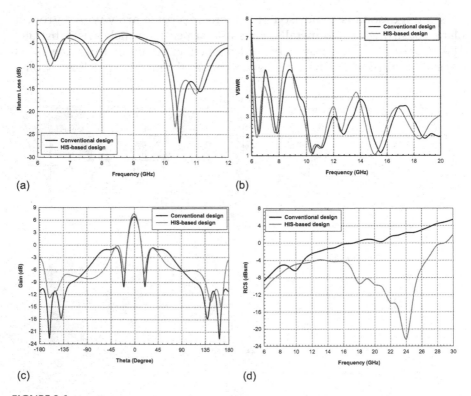

FIGURE 3.6
Radiation and scattering characteristics of a 4-element patch array with non-uniform HIS layer (**Configuration 2: Three-layered**) and modified ground plane at 10.5 GHz (substrate: **60 × 90 mm**): (a) Return loss, (b) VSWR, (c) gain, and (d) specular structural RCS.

The corresponding values of VSWR are 1.1 and 1.15, respectively (Figure 3.6b). Figure 3.6c shows that the gain of the patch array has improved from 6.75 to 7.46 dB. The structural RCS plot (Figure 3.6d) shows that this configuration of HIS provides significant out-of-band RCS reduction. The extent of RCS reduction when compared to the patch array with conventional ground plane is shown as a bar chart in Figure 3.7a. RCS reduction of almost 2 dBsm is observed from 6 to 9 GHz. It is to be noted that the maximum RCS reduction of 24.72 dBsm is achieved at 24 GHz. Thus, this configuration is found to be efficient in terms of RCS reduction with an improved gain. The band-stop behavior of the HIS structure can be observed from Figure 3.7a; the array RCS has increased by 1.5 dBsm at 10 GHz, close to resonant frequency. Figure 3.7b shows the structural RCS of the patch array at 24 GHz, where maximum specular RCS reduction has been obtained. It can be observed that in case of hybrid HIS-based design, null is obtained at the specular lobe. In other words, RCS reduction of 24.97 dBsm is observed at the specular lobe.

Further, the dimension of the antenna along the length was reduced from 60 to 45 mm and a similar HIS configuration has been implemented on the patch array. Figure 3.8 depicts the top, middle and bottom layers of the patch array with hybrid HIS layer and modified ground plane. The radiation and scattering characteristics of the designed array are shown in Figure 3.9. It can be observed from Figure 3.9a that the patch array with HIS-based ground plane resonates at 10.33 GHz with a return loss of −23.28 dB. The radiating patch array has a standing wave ratio of 1.15 (Figure 3.9b) at the resonating frequency. Figure 3.9c shows that there is no degradation in the gain with HIS-based design. The patch array with conventional ground plane shows a gain of 7.3 dB, while the patch array based on hybrid HIS layer and modified

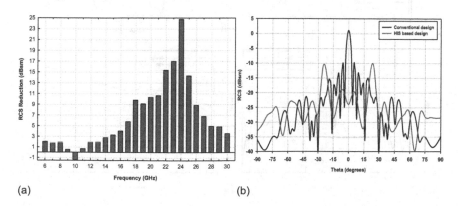

(a) (b)

FIGURE 3.7
(a) Variation in specular RCS reduction with frequency and (b) structural RCS pattern at 24 GHz.

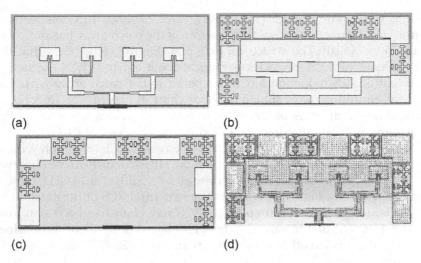

FIGURE 3.8
Three-layered design: A 4-element linear patch array with non-uniform HIS layer with modified HIS ground plane at 10.5 GHz (Substrate: **45 × 90 mm**): (a) Top layer, (b) middle layer, (c) bottom layer, and (d) surface current distribution at resonant frequency.

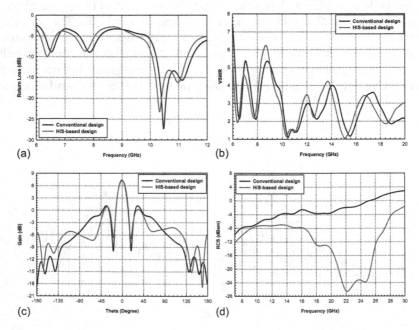

FIGURE 3.9
Radiation and scattering characteristics of a 4-element patch array with hybrid HIS layer and modified ground plane at 10.5 GHz (Substrate: **45 × 90 mm**): (a) Return loss, (b) VSWR, (c) gain, (d) specular structural RCS.   (*Continued*)

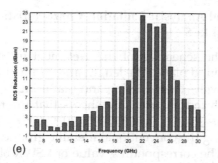

(e)

FIGURE 3.9 (Continued)
(e) variation in specular RCS reduction with frequency.

ground plane shows a gain of 7.38 dB. Thus, the radiation performance of the patch array has not degraded after modifying the ground plane and the inclusion of the HIS layer.

The scattering characteristics of the patch array are shown in Figure 3.9d. It can be observed that a wide-band RCS reduction is achieved for the frequency range of 7–30 GHz. The maximum RCS reduction (24.5 dBsm) is achieved at 22 GHz.

Figure 3.9e shows the variation in specular RCS reduction with frequency. It can be observed that as the antenna array size is reduced to 45 × 90 mm, RCS reduction around the in-band frequency range (10.5 GHz) is also achieved.

In this case, the results obtained are satisfactory in view of wideband RCS reduction. As mentioned above, RCS reduction is achieved for both in-band and out-of-band frequency range. However, there are fabrication related challenges in case of this three-layered geometry. Thus, the next objective is to achieve similar or better antenna performance with a two-layered structure. The next section deals with design and analysis of such two-layered microstrip patch array with hybrid HIS layer as ground plane.

3.3 Patch Array with Hybrid Ground Plane: Two-Layered Geometry

In this two-layered design configuration, the conventional metallic ground plane of the patch array is replaced by a hybrid HIS layer, mentioned in Figure 3.5b. The top layer is the patch array with corporate feed and the second layer is HIS-based ground plane. The objective toward the design of the two-layered structure was to reduce the complexity of the design and thus minimize simulation and fabrication time.

3.3.1 Simulation and Analysis: Radiation Characteristics and Structural RCS

Figure 3.10 shows the schematic of the 4-element microstrip patch array with HIS layer as ground plane. The length and width of the substrate are 60 and 90 mm, respectively. The radiation and scattering behavior of designed patch array are analyzed. Figure 3.11a shows that the resonant frequency of the patch array has shifted from 10.48 to 10.37 GHz with a return loss of −29.24 dB, by replacing the conventional metallic ground plane by modified HIS ground plane. The corresponding value of VSWR obtained is 1.16. It is apparent from Figure 3.11b that there is an improvement in the array gain.

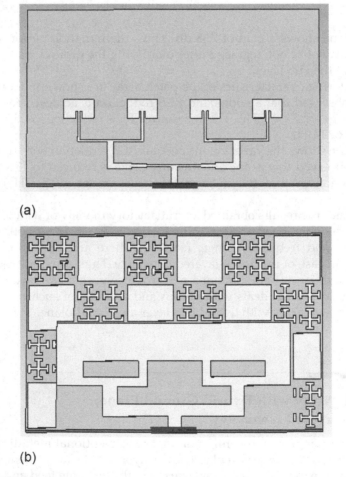

(a)

(b)

FIGURE 3.10

Two-layered design: A 4-element linear patch array with hybrid HIS layer as ground plane at 10.5 GHz (Substrate dimensions: **60 × 90 mm**): (a) Top view and (b) bottom view.

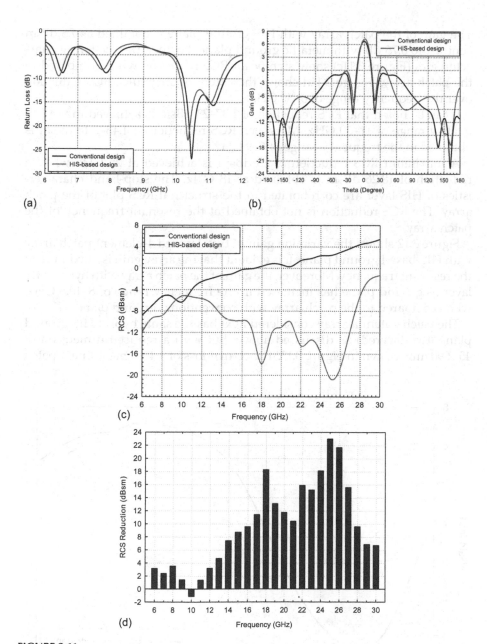

FIGURE 3.11
Radiation and scattering characteristics of a two-layered 4-element patch array with modified ground plane: (a) Return loss, (b) gain, (c) specular structural RCS, and (d) variation in specular RCS reduction with frequency.

The gain of the patch array with HIS ground plane is obtained as 7.3 dB in contrast to 6.75 dB with metallic ground plane. Further, the scattering characteristics (Figure 3.11c) show significant RCS reduction when compared to the patch array. Figure 3.11d shows the extent of structural RCS reduction over the frequency range of 6–30 GHz.

It can be noted that RCS reduction of 1.37–23 dBsm is achieved in the frequency range of 11–30 GHz. Moreover, RCS reduction of 1.43–3.16 dBsm is obtained in the frequency range of 6–9 GHz.

A maximum RCS reduction of 23 dBsm is achieved at 25 GHz while a reduction of 18.26 dBsm is obtained at 18 GHz. The stop-band characteristics of HIS layer are corroborated in the structural RCS plot of the patch array. The RCS reduction is not obtained at the resonant frequency of the patch array.

Figure 3.12 shows the variation gain of two-layered 4-element patch array with HIS-based ground plane. It is evident that the array gain is maximum at the resonant frequency. Moreover, the gain of the patch array with hybrid HIS layer as ground plane has increased in the frequency range of 8–10.5 GHz, when compared to the patch array with conventional ground plane.

The patch antenna array considered next has the same hybrid HIS ground plane (two-layered) as discussed above, but with a reduced dimension of 45 × 90 mm shown in Figure 3.13. The return-loss of a 4-element linear patch

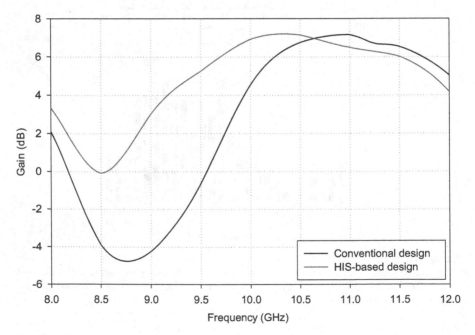

FIGURE 3.12
Variation of gain of 4-element patch array with HIS-based ground plane.

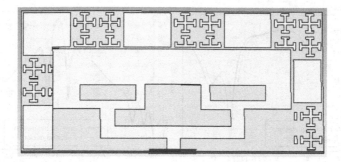

FIGURE 3.13
Two-layered design: Ground plane of the 4-element linear patch array with hybrid HIS layer (Substrate dimension: **45 × 90 mm**).

array with conventional ground plane (45 × 90 mm) achieved is –27.14 dB at 10.46 GHz, as shown in Figure 3.14a. On the other hand, the return-loss of the patch array with hybrid HIS layer as ground plane is obtained as –22.97 dB at 10.35 GHz. The VSWR of the patch array has slightly increased from 1.09 to 1.15. It can be noted from Figure 3.14b that the gain of both patch array, one with conventional ground plane and another with hybrid HIS layer as ground plane is the same, i.e., 7.3 dB. The scattering characteristics of the antenna array with modified ground plane are shown in Figure 3.14c and d, respectively.

The structural RCS pattern of the reference and proposed antenna is compared in Figure 3.14c along the elevation plane ranging from –90° to 90°. The RCS pattern of the patch array with conventional ground plane is found to be maximum along the specular range with a value of –7.95 dBsm. Further, a reduction of 1.93 dBsm can be observed for patch array with modified ground plane. It can be noted that the same trend is observed in scattering behavior when the size of the patch array is reduced from 60 × 90 mm to 45 × 90 mm. When the size of the patch array is reduced to 45 × 90 mm, both in-band and out-of-band RCS reduction is observed.

When compared to the three-layered patch array structure, the results have improved for the two-layered patch array design. It is apparent from Figure 3.14d that RCS reduction has been achieved for the entire range of 1–80 GHz and further with a minimum and maximum reduction of 1.82 and 27.07 dBsm, respectively. Thus, one can conclude that wideband structural RCS reduction has been achieved with the help of two-layered design of patch array with HIS-based hybrid ground plane.

The variation of gain of the designed two-layered patch array with HIS-based ground plane is shown in Figure 3.15. It can be noted that by replacing the conventional metallic ground plane of the patch array with hybrid HIS layer, the enhancement in the array gain is achieved. However, for frequencies higher than the resonant frequency, the gain has reduced.

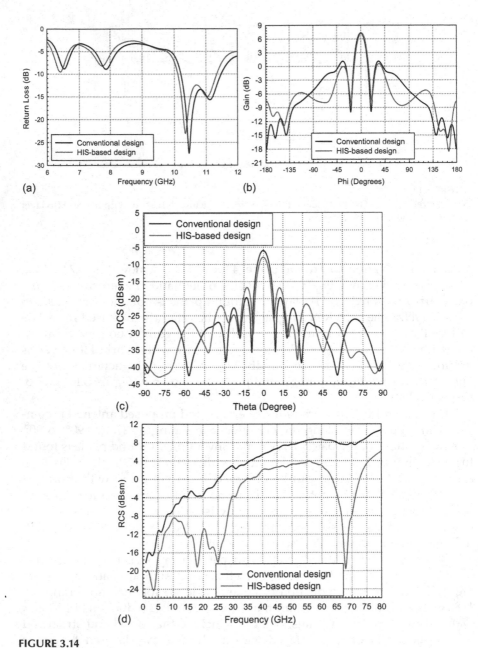

FIGURE 3.14
Radiation and scattering characteristics of a 4-element patch array with modified ground plane: (a) Return loss, (b) gain, (c) structural RCS pattern at 10.46 GHz, and (d) specular structural RCS with respect to frequency.

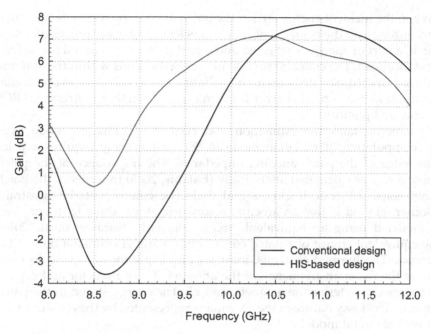

FIGURE 3.15
Variation of gain with frequency.

3.3.2 Scattering Characteristics: Radiation Mode RCS

It has been well-established that high impedance surface (HIS) plays an important role in controlling the radiation and scattering characteristics of an antenna array. As discussed in the previous section, a four element linear patch array with hybrid HIS-based ground plane was analyzed for structural mode radar cross section (RCS). It was shown that the HIS layer was able to reduce the structural RCS for both in-band and out-of-band frequencies. As mentioned earlier, for a radiating structure such as a patch array, the radiation mode RCS plays a significant role when compared to the structural mode RCS in the total array RCS. It is due to the fact that the structural RCS of an antenna array is computed when the antenna impedance is perfectly matched to the feed structure. To achieve a perfectly matched state is practically not possible.

The corporate feed structure in the antenna array itself contributes to most of the mismatches. The corporate feed network consists of various levels where each level has a feedline of different feed width. Each feedline of a certain feed width acts as a transmission line having a certain input impedance and characteristic impedance. Apart from the feed structure, the feed network is comprised of couplers and phase-shifters. Thus, the impedance mismatches between the coupler and the feed network and in between the phase-shifter and the feed network also result in reflections.

When the incident energy strikes the patch array, it traverses the feed network through all the components at various levels. At every junction of mismatch, a part of energy is reflected and a part of it is transmitted to the feed network ahead. The coherent sum of all these reflections within the antenna and feed structure contributes to the radiation mode RCS. The level of radiation mode RCS of a phased array is much higher than that of structural RCS of array and platform.

This necessitates the estimation and control of radiation mode RCS. The computation of radiation mode RCS of patch array requires a prior knowledge of the patch antenna impedance. The impedance of the patch antenna can be calculated analytically (Balanis, 2005) in terms of the patch dimensions, substrate thickness and dielectric constant of the substrate. Another method is that an antenna resonating at a specific frequency can be modeled using an equivalent circuit approach (Narayan et al., 2016). The equivalent circuit of a patch consists of a parallel combination of RLC network. The lumped elements, inductor and capacitor are tuned to resonate at the resonant frequency of the antenna. These inductor and capacitor values characterize the magnetic and electric energy stored in the patch antenna. The lossy nature of the antenna is represented by the resistor in the equivalent circuit model.

The HIS layer needs to be modeled separately for an antenna array. The HIS-based ground plane consists of a uniform layer of FSS elements. The researchers have reported techniques (Costa et al., 2014) to represent the FSS and meta-surfaces by their equivalent circuit model. These models are limited to simple FSS configurations. The equivalent circuit of non-resonant elements such as a patch, strip or a wire grid can be modeled by a capacitor. Single resonant structures such as a loop or a dipole cross can be modeled by a single LC combination. Double resonant structures, as in our case the Jerusalem cross, can be modeled by a parallel combination of two LC elements. However, for multi-resonant structures it is physically not possible to design the equivalent circuit model. Furthermore, for complex structures consisting of two AMC elements with a hybrid ground plane, it is not possible to compute the antenna impedance from equivalent circuit approach. Due to the presence of a non-uniform HIS layer beneath the patch antenna, the value of antenna impedance will vary.

Let us consider a 4-element patch array with HIS-based ground plane (Figure 3.10). It is evident from the patch array structure that the HIS pattern that comes beneath each patch in the array is different. This implies that the corresponding antenna impedance will also be different for each patch.

Figure 3.16 shows the top and bottom layer of the single patch antennas made from dividing the patch array into four segments. Each antenna is separately fed through a microstrip feedline, and resonates at 10 GHz.

The return loss, antenna impedance, and directivity of the four patch antenna segments obtained are listed in Table 3.1. The radiation mode RCS of the 4-element microstrip patch array with hybrid HIS-based ground plane

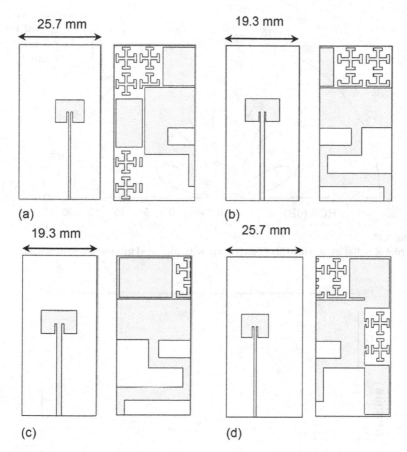

FIGURE 3.16
Antenna segments of a 4-element linear patch array with hybrid HIS layer as ground plane for impedance calculation. (a) Segment 1, (b) Segment 2, (c) Segment 3, and (d) Segment 4.

TABLE 3.1

Radiation Characteristics of the Antenna Segments of a 4-Element Patch Array with Hybrid HIS-Based Ground Plane

Antenna Segment	Return Loss	Antenna Input Impedance	Directivity
Segment 1	−31.85 dB at 9.99 GHz	$191.5 - j\,8.5\ \Omega$	5.02
Segment 2	−25.1 dB at 9.99 GHz	$184.8 - j\,11.5\ \Omega$	1.62
Segment 3	−35.9 dB at 9.99 GHz	$142.16 - j\,3.92\ \Omega$	2.32
Segment 4	−37.2 dB at 10.048 GHz	$187.3 + j\,8.96\ \Omega$	3.51

is computed. Figure 3.17 shows the radiation mode RCS at 10 GHz for different values of characteristic impedance, with load impedance fixed at 75 Ω. It can be observed that the specular RCS for Z_0 values of 30, 50 and 75 Ω are respectively, 15.06, 14.04 and 12.5 dB.

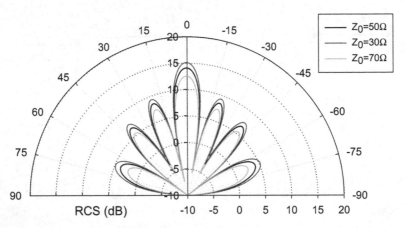

FIGURE 3.17
Polar plot of radiation mode RCS of patch array with modified ground plane at 10 GHz.

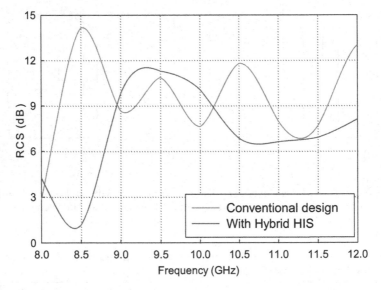

FIGURE 3.18
Radiation mode RCS of 4-element patch array with modified HIS-based ground plane, shown in Figure 3.13.

Figure 3.18 shows the analysis done for a 4-element linear patch array with conventional ground plane and hybrid HIS-based ground plane for characteristic and load impedance values of 150 and 60 Ω, respectively. It can be observed that the radiation mode RCS for conventional ground plane is 7.61 dB at 10 GHz, while for the patch array with hybrid HIS-based ground plane is 10.07 dB. It can be noted that for out-of-band frequency ranges, the patch array with hybrid HIS-based ground plane is able to reduce the specular radiation mode RCS when compared to patch array with conventional metallic ground plane.

3.4 Summary

Artificial magnetic conductor (AMC) elements are capable of reducing structural RCS of microstrip patch arrays. The bandwidth and the extent of RCS reduction they can provide depend on the type of AMC element chosen, its resonant frequency and other design parameters, and also the configuration designed. A combination of AMC cells and square patches in chessboard configuration provide narrowband RCS reduction through phase cancellation of reflected waves. However, by implementing a hybrid chessboard configuration of AMC-square patch on the ground plane of an antenna/array, wide band reduction can be achieved. In this chapter, the radiation and scattering analysis of a microstrip patch antenna/array with hybrid HIS-based ground plane has been presented. Wideband RCS reduction was achieved for both two-layered and three-layered design configuration. It can be concluded that the scattering behavior of a microstrip patch array can be varied, without affecting the radiation characteristics, if the ground plane is replaced with a suitable configuration of AMC elements. Minimum radiation mode RCS can be obtained even in the in-band frequency region, when the antenna impedance is matched to the characteristic and load impedances. Further, this results in the degradation of the antenna performance such as the gain. Thus, there is always a trade-off between gain and array RCS with appropriate choice of characteristic and load impedances. The RCS analyses done shows that the patch array with hybrid HIS-based ground plane provides specular RCS reduction in the out-of-band frequency range with an optimum RCS in the in-band frequency range, when compared to patch arrays with conventional metallic ground plane for the same values of characteristic and load impedances. Therefore, proper selection of characteristic and load impedances is one of the necessary steps toward optimizing radiation mode RCS along with desired radiation performance.

4

Low RCS Conformal Array and Effect of Hybrid Ground Plane

The property of high impedance surfaces (HIS) to control the scattering of EM wave has been widely used in attaining reduction in radar cross section (RCS) of microstrip patch arrays. It has been reported in the previous chapter that incorporating a hybrid HIS-based ground plane in a 4-element planar microstrip patch array reduces the structural RCS over a wide frequency range. Further low profile patch array over a curved surface contributes in reducing the antenna scattering (Singh & Singh, 2018). It is well known that the conformal patch arrays are the preferred choice owing to their conformal geometry reducing the protruding structures from the platform. Furthermore, as compared to planar patch array, a conformal patch array scatters the incident EM wave in the directions other than specular one, thereby reducing the array RCS (Josefsson & Persson, 2006). In addition, the inclusion of HIS-based hybrid ground plane further reduces the array RCS.

The radiation characteristics of a microstrip patch array depend on the substrate material as well. Low loss dielectric materials can be used as the substrate of patch arrays, since they are capable of providing larger gain efficiency. Further, thick substrates with low dielectric constant can provide larger bandwidth (Balanis, 2005). This chapter pertains to the EM design and analysis of microstrip patch array with hybrid HIS-based ground plane on a low loss RT Duroid substrate. Further, FR4 substrate material is brittle in nature and RT Duroid when compared to FR4 can be conformed on a curved platform for a certain radius of curvature. The radiation and scattering analysis is presented for both planar as well as conformal patch arrays.

4.1 Planar Patch Array with Full Ground Plane

This section deals with the EM design and analysis of microstrip patch array with conventional ground plane. The 4-element linear patch array is designed to resonate in the X-band. A low-loss dielectric, Rogers RT Duroid (5880) is used as the substrate material. The substrate has a thickness of 1.58 mm with relative permittivity (ε_r) and loss tangent (tan δ) as 2.2 and

0.0009 respectively. The patch is made of copper material with a conductivity of 5.8×107 S/m. The patch element has a dimension of 9.07×11.86 mm^2 with a thickness of 0.018 mm. An inter-element spacing of 0.643λ is considered between patch elements. The patch array has a dimension of 49×90 mm. Figure 4.1 shows the schematic of the 4-element planar patch array along with a corporate feed structure, fed by a waveguide port. The 50 and 100 Ω feedline has a width of 3.4 and 1.3 mm, respectively. The stub line (70.7 Ω) used to match the 50 and 100 Ω feedlines has a feed width of 2.4 mm.

The radiation and scattering characteristics of the patch array have been evaluated using a full-wave simulation software.

Figure 4.2 shows the radiation characteristics of the planar patch array with conventional ground plane. It can be observed from Figure 4.2a that the patch array resonates at 9.78 GHz with a return loss of −26.53 dB.

The planar patch array with conventional ground plane has a bandwidth of 9.58%. The VSWR of the radiating patch array has a value of 1.09. The planar patch array has a gain of 12.3 dB along the specular direction, i.e., $\theta = 0°$ (Figure 4.2b) with a side lobe level of −7.9 dB.

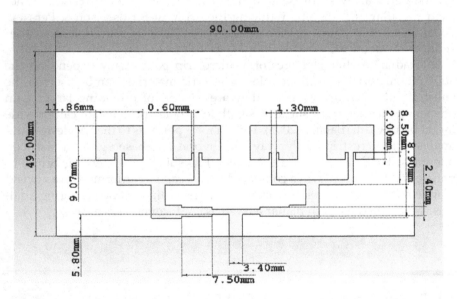

FIGURE 4.1
Planar patch array with conventional metallic ground plane.

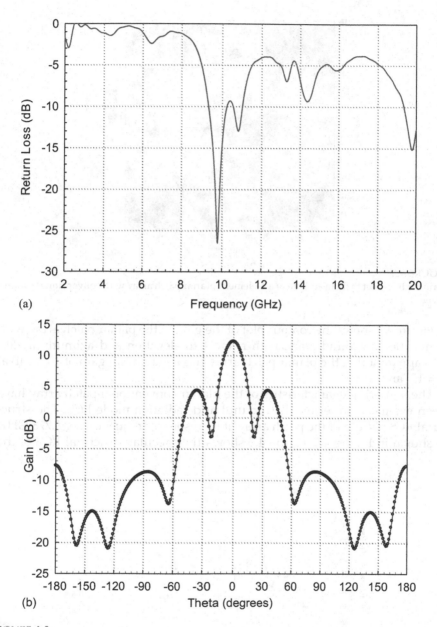

(a)

(b)

FIGURE 4.2

Radiation characteristics of a 4-element **planar** patch array with **conventional** ground plane: (a) Return loss and (b) gain ($\phi = 90°$).

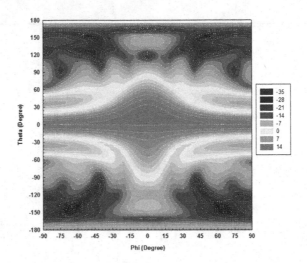

FIGURE 4.3
Gain with respect to theta and phi of a 4-element **planar** patch array with **conventional** ground plane.

Figure 4.3 shows the contour plot of the gain of the planar patch array with conventional ground plane with respect to elevation and azimuth angles. It is apparent that the planar patch array has a maximum gain of 13.33 dB at $\theta = 15°$ and $\phi = 0°$.

The scattering characteristics of the planar conventional patch array have been evaluated in terms of structural and radiation mode RCS. The structural RCS pattern of the patch array at its resonant frequency, i.e., 9.778 GHz, is shown in Figure 4.4. Figure 4.5 shows the specular structural RCS of the

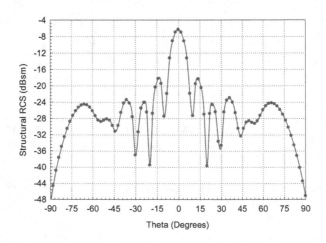

FIGURE 4.4
Structural RCS pattern of 4-element planar patch array with **conventional** metallic ground plane at the resonant frequency.

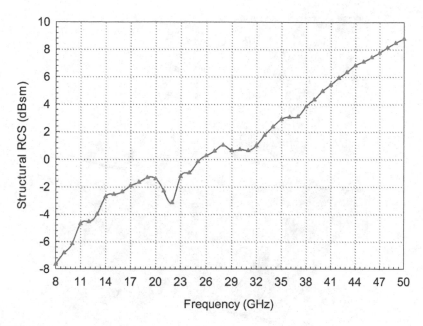

FIGURE 4.5
Specular structural RCS pattern of 4-element planar patch array with **conventional** metallic ground plane with respect to frequency.

planar patch array with conventional ground plane from 8 to 50 GHz. It can be noted that the specular structural RCS increases almost linearly with the increase in frequency.

Radiation Mode RCS: The patch array is divided into four segments as shown in Figure 4.6. Each patch segment is fed by a microstrip feed line. The impedance and the directivity of each patch segment are obtained at resonant frequency of 9.77 GHz (Table 4.1) and other frequency, i.e., 10 GHz (Table 4.2). The real part of the impedance, i.e., the resistance corresponds to the radiated power while the imaginary part, i.e., the reactance relates with the non-radiated power in the near field region. Thus for maximum radiation, reactance of the antenna should be minimum. It can be observed that Table 4.1 shows higher reactance and less resistance as compared to Table 4.2. Thus, the antenna segments at 10 GHz has more directivity than at 9.77 GHz.

Figure 4.7a and b show the radiation mode RCS of the patch array at 9.77 and 10 GHz for different values of load impedances. The characteristic impedance (Z_o) is fixed at 50 Ω. The specular RCS value is obtained as 13.5 and 16.05 dB for load impedance (Z_L) of 75 and 50 Ω, respectively. The variation of radiation mode RCS with frequency is shown in Figure 4.7c. It is evident that the array RCS is maximum near the resonant frequency, which is as per expectation lines.

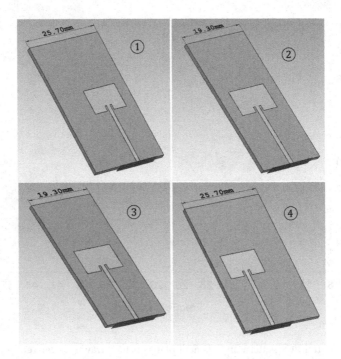

FIGURE 4.6
Antenna segments of a 4-element planar patch array with conventional metallic ground plane.

TABLE 4.1

Radiation Characteristics of the Antenna
Segments of a 4-Element Planar Patch Array with
Conventional Metallic Ground Plane at 9.77 GHz

Antenna Segment	Antenna Impedance	Directivity
Segment 1	$95.73 + j\,61.07\ \Omega$	3.91
Segment 2	$95.34 + j\,62.95\ \Omega$	3.63
Segment 3	$95.34 + j\,62.95\ \Omega$	3.63
Segment 4	$95.73 + j\,61.07\ \Omega$	3.91

TABLE 4.2

Radiation Characteristics of the Antenna
Segments of a 4-Element Planar Patch Array with
Conventional Metallic Ground Plane at 10 GHz

Antenna Segment	Antenna Impedance	Directivity
Segment 1	$138.96 + j\,20.6\ \Omega$	4.41
Segment 2	$140.92 + j\,22.06\ \Omega$	4.06
Segment 3	$140.9 + j\,22.08\ \Omega$	4.06
Segment 4	$138.96 + j\,20.6\ \Omega$	4.41

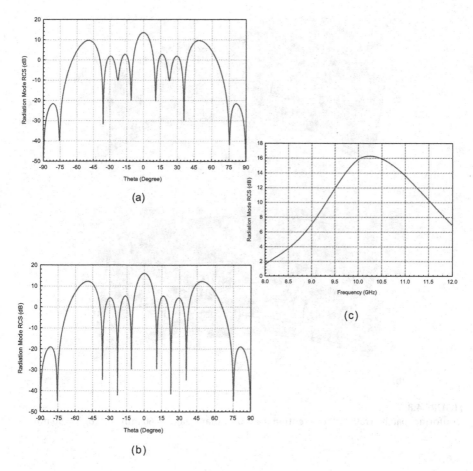

FIGURE 4.7
Radiation mode RCS of a 4-element **planar** patch array with **conventional** ground plane: (a) f_o = 9.7 GHz, Z_o = 50 Ω, Z_L = 75 Ω; (b) f_o = 10 GHz, Z_o = 50 Ω, Z_L = 50 Ω; and (c) RCS with respect to frequency.

4.2 Conformal Patch Array with Full Ground Plane

The planar patch array discussed in the previous section (Figure 4.1) is conformed to a cylindrical surface. Figure 4.8a and b show the conformal patch array with conventional metallic ground plane. Two cases with different radii of curvature 60 and 90 mm have been considered. The patch array has the same dimensions as mentioned for a planar patch array.

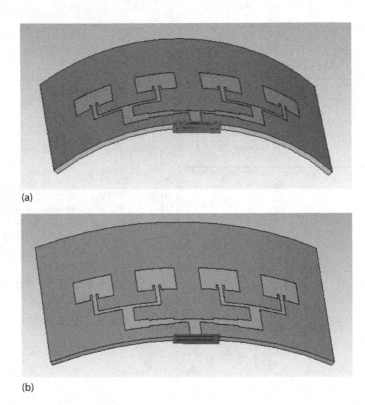

(a)

(b)

FIGURE 4.8
Conformal patch array with **conventional** ground plane: (a) $R = 60$ mm and (b) $R = 90$ mm.

4.2.1 Radius of Curvature = 60 mm

The radiation and scattering characteristics of the conformal patch array
with conventional ground plane and radius of curvature as 60 mm has
been analyzed. It can be observed from Figure 4.9a that the conformal
patch array resonates at 9.51 GHz with a return loss of −37.31 dB. It has
a percentage bandwidth of 9.52 in the X-band. The radiating conformal
patch array has a VSWR of 1.02. It can be noted from Figure 4.9b that the
gain of the conformal patch array along the specular direction is reduced
from 12.3 dB (Figure 4.2b) to 10.01 dB when compared to the planar patch
array.

The reduction in gain can be attributed to the curved structure of the array.
A contour plot illustrating the gain with respect to elevation and azimuth
plane of the conformal patch array with conventional ground plane having
radius of curvature 60 mm is shown in Figure 4.10. It can be noted that the
maximum gain of 11.4 dB of the conformal patch array is obtained along
$\theta = 20°$ and $\phi = 0°$.

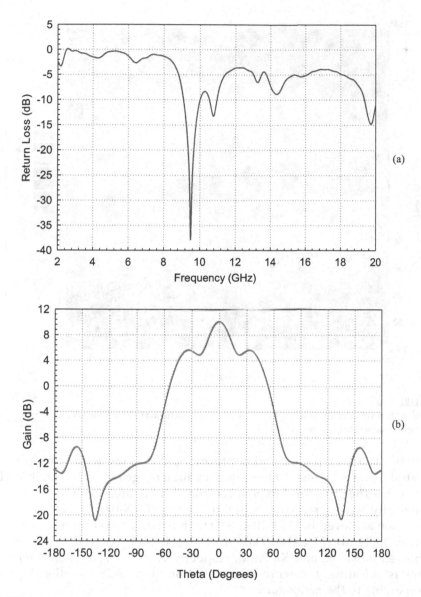

FIGURE 4.9
Radiation characteristics of conformal patch array with **conventional** ground plane: (a) Return loss and (b) gain.

The scattering analysis of the conformal patch array with conventional ground plane is done in terms of its structural RCS. Figure 4.11a represents the structural RCS pattern at the resonant frequency, i.e., 9.51 GHz. It can be noted that the structural RCS does not have a peak in the specular direction. Figure 4.11b shows the comparison of specular

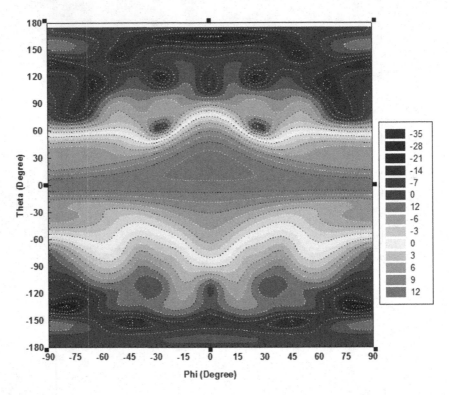

FIGURE 4.10
Gain with respect to theta and phi of a 4-element **conformal (R = 60 mm)** patch array with **conventional** ground plane.

structural RCS of planar and conformal patch array with conventional ground plane with respect to frequency for a range of 8–50 GHz. It can be noted that the structural RCS of the conformal patch array has reduced drastically when compared to planar array. In X-band, the maximum reduction achieved is 13.74 dB at 8 GHz. Further, a maximum RCS reduction of 15.34 dB (15 GHz), 13.08 dB (23 GHz), and 19.10 dB (37 GHz) is achieved in Ku, K, and Ka band, respectively. However, when the patch array is radiating, the array RCS (radiation mode RCS) is either high or comparable to the array gain.

This high value of array RCS necessitates its accurate estimation and hence, its control, without degrading the radiation performance. In order to compute the radiation mode RCS, the conformal patch array is divided into antenna segments as shown in Figure 4.12. The impedance and the directivity of each patch segment are obtained at the resonant frequency (9.51 GHz), and at 10 GHz as mentioned in Tables 4.3 and 4.4, respectively.

Figure 4.13a and b show the radiation mode RCS of the patch array at 9.51 and 10 GHz for characteristic (Z_o) and load impedance (Z_L) values of 50 and 75 Ω, respectively.

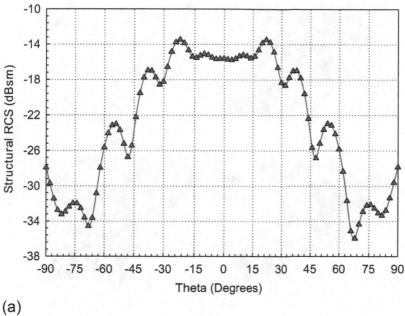

(a)

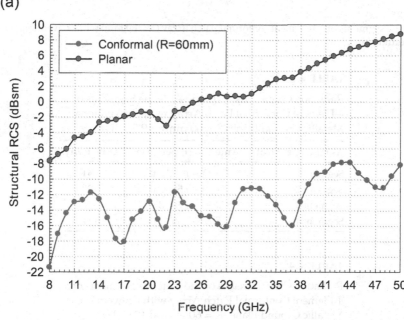

(b)

FIGURE 4.11

Scattering characteristics of conformal patch array with **conventional** ground plane: (a) Return loss and (b) gain.

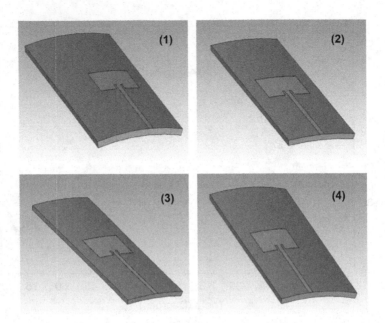

FIGURE 4.12
Antenna segments of a 4-element conformal patch array ($R = 60$ mm) with conventional metallic ground plane.

TABLE 4.3

Radiation Characteristics of the Antenna Segments of a
4-Element Conformal Patch Array with Conventional
Metallic Ground Plane ($R = 60$ mm) at 9.51 GHz

Antenna Segment	Antenna Impedance	Directivity
Segment 1	$75.73 + j\,55.37\ \Omega$	3.41
Segment 2	$77.19 + j\,58.23\ \Omega$	3.20
Segment 3	$77.19 + j\,58.23\ \Omega$	3.20
Segment 4	$75.77 + j\,55.32\ \Omega$	3.42

TABLE 4.4

Radiation Characteristics of the Antenna Segments of a
4-Element Conformal Patch Array with Conventional
Metallic Ground Plane ($R = 60$ mm) at 10 GHz

Antenna Segment	Antenna Impedance	Directivity
Segment 1	$133.61 - j\,33.14\ \Omega$	4.47
Segment 2	$141.73 - j\,42.47\ \Omega$	4.19
Segment 3	$141.73 - j\,42.47\ \Omega$	4.19
Segment 4	$133.55 - j\,33.19\ \Omega$	4.47

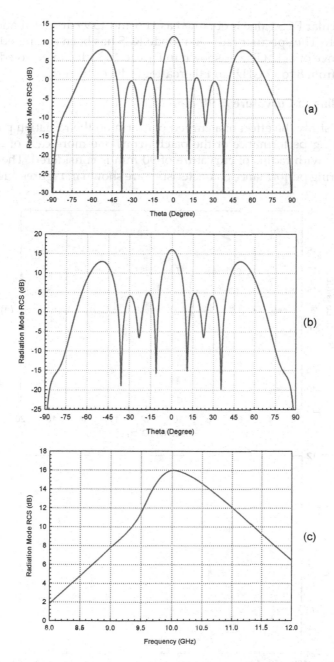

FIGURE 4.13
Radiation mode RCS of a 4-element **conformal** patch array with **conventional** ground plane ($R = 60$ mm) at $Z_o = 50$ Ω, $Z_L = 75$ Ω: (a) $f = 9.51$ GHz, (b) $f = 10$ GHz, and (c) RCS with respect to frequency.

The specular RCS value is obtained as 11.6 and 15.99 dB at 9.51 and 10 GHz, respectively. The specular radiation mode RCS has been computed for a frequency range of 8–12 GHz as shown in Figure 4.13c. It can be noted that RCS increases from 8 to 10 GHz, and thereafter reduces linearly.

4.2.2 Radius of Curvature = 90 mm

To further study the effect of a curved platform on the radiation pattern and the scattering performance of the patch array, one more case of conformal patch array with radius of curvature of 90 mm is considered. The radiation and scattering performance characteristics are shown in Figures 4.14 and 4.15.

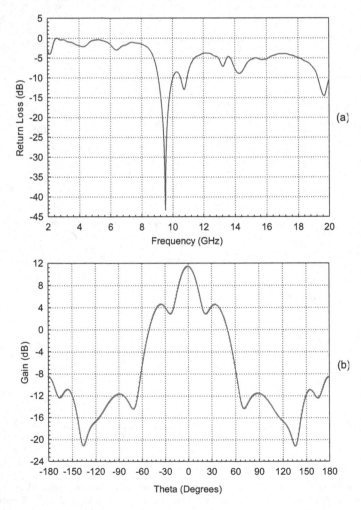

FIGURE 4.14
Radiation characteristics of conformal patch array (R = 90 mm) with **conventional** ground plane: (a) Return loss and (b) gain.

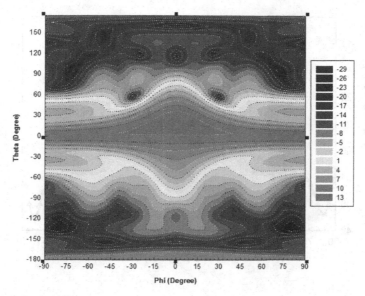

FIGURE 4.15
Gain with respect to theta and phi of a 4-element **conformal** (*R* = 90 mm) patch array with
conventional ground plane.

It can be observed from Figure 4.14a that the conformal patch array has a
return loss of −42.48 dB at 9.53 GHz with a bandwidth of 9.88%. The VSWR
obtained is 1.04 at the resonant frequency. It can be noted that the gain of
the conformal patch array with radius of curvature 90 mm is slightly greater
than that of patch array with radius of curvature of 60 mm. Figure 4.14b
shows the gain obtained along the specular direction is 11.36 dB. The contour plot of the gain of the conformal patch array with conventional ground
plane is illustrated in Figure 4.15. It can be observed that the maximum gain
(12.5 dB) is obtained along the direction $\theta = 20°$ and $\phi = 0°$. It can be observed
from Figures 4.3, 4.10 and 4.15 that the gain of the conformal patch array is
prominent in directions other than the specular direction as well.

The scattering characteristics of the conformal patch array with conventional ground plane with radius of curvature 90 mm have been analyzed.
Figure 4.16a and b represents the structural RCS of the conformal patch
array. It can be noted that the structural RCS has a null in the specular direction with a value of −16.3 dBsm (Figure 4.16a). The specular structural RCS of
the conformal patch array has also been computed with respect to frequency
from 8 to 50 GHz. It can be observed from Figure 4.16b that unlike the planar patch array, the specular structural RCS of the conformal patch array
does not increase linearly with frequency. A minimum reduction of 8 dB
and a maximum reduction of 20 dB in specular structural RCS of conformal
patch array with conventional ground plane has been observed. In planar
patch array, the surface normal of each element points in the same direction

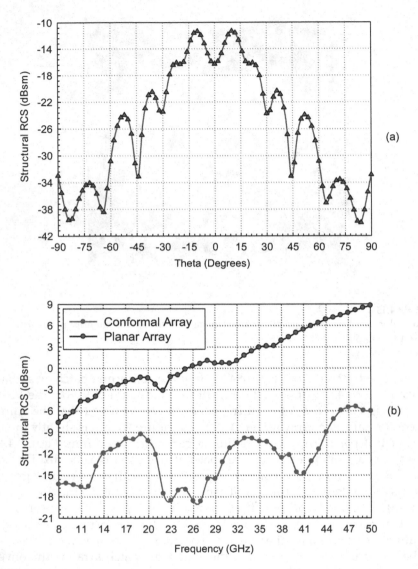

FIGURE 4.16

Scattering characteristics of conformal patch array (R = 90 mm) with conventional ground plane: (a) Structural RCS pattern and (b) specular structural RCS with respect to frequency.

(specular) while in conformal patch array the surface normal of each element points in a different direction. The array RCS is the coherent sum of all these points which results in reduced RCS for conformal patch array.

For radiation mode RCS, the conformal patch array is divided into antenna segments. The impedance and the directivity of each segment is given in Table 4.5. Figure 4.17a shows the radiation mode RCS of the conformal patch array with conventional ground plane at 10 GHz. It can be observed that the

TABLE 4.5

Radiation Characteristics of the Antenna Segments of a
4-Element Conformal Patch Array with Conventional
Metallic Ground Plane ($R = 90$ mm) at 10 GHz

Antenna Segment	Antenna Impedance	Directivity
Segment 1	$137.1 - j\,35.5\,\Omega$	4.58
Segment 2	$147.2 - j\,34.6\,\Omega$	4.25
Segment 3	$147.2 - j\,34.6\,\Omega$	4.25
Segment 4	$137.1 - j\,35.4\,\Omega$	4.58

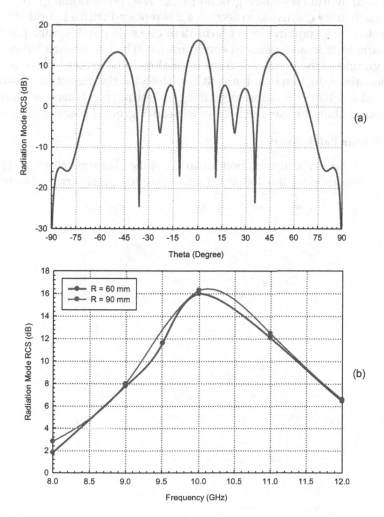

(a)

(b)

FIGURE 4.17
Radiation mode RCS of 4-element conformal patch array with conventional ground plane;
$Z_o = 50\,\Omega$, $Z_L = 75\,\Omega$: (a) $f = 10$ GHz and (b) specular RCS with respect to frequency.

specular lobe has a value of 16.29 dB. Figure 4.17b illustrates the comparison between the radiation mode RCS with respect to frequency in the X-band. It is noted that the RCS of the patch array increases with the increase in the radius of curvature of the curved platform, which approximates toward the planar array.

4.3 Conformal Patch Array with Hybrid Ground Plane

This section deals with the EM design and analysis of 4-element linear patch arrays with hybrid HIS-based ground plane. The conventional ground plane of the patch array discussed in Section 4.2 is replaced with a hybrid HIS layer constituted by a combination of Jerusalem cross (JC) and square patch elements arranged in a chessboard configuration. The use of such hybrid HIS-based ground plane instead of conventional metallic ground plane resulted in substantial reduction in antenna RCS with FR4 as the dielectric substrate as discussed in Chapter 3. Before describing conformal patch array with hybrid HIS ground plane, planar patch array with such ground plane is discussed.

4.3.1 Planar Patch Array

Figure 4.18 shows the top and bottom layers of the planar patch array. The top layer is comprised of the 4-element corporate fed patch array. The bottom

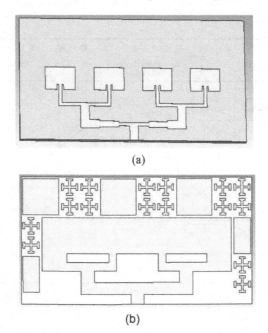

(a)

(b)

FIGURE 4.18
Planar patch array with hybrid HIS-based ground plane: (a) Top layer and (b) bottom layer.

layer includes the hybrid HIS layer, comprising of artificial magnetic conductor (AMC) elements and a metallic patch with the outline of the patch array designed on the top.

The planar patch array shown in Figure 4.18 has RT Duroid as the dielectric substrate. There are mainly two advantages of RT Duroid over FR4. Firstly, RT Duroid (tan δ = 0.0009) is low loss dielectric as compared to FR4 (tan δ = 0.022). Moreover, RT Duroid has flexible surface. Thus, the design results are expected to be better than that of FR4 substrate. Secondly, patch array based on RT Duroid substrate can be made conformal which is not possible for patch array with FR4 as substrate. The planar patch array with hybrid HIS-based ground plane has the same antenna configuration as discussed in Chapter 3.

The radiation characteristics of the planar patch array are shown in Figure 4.19. It can be noted from that the array is resonating at 9.78 GHz with a return loss of −39.35 dB. When compared to the planar patch array with conventional ground plane, the patch array with hybrid HIS-based ground plane is resonating at the same frequency with an improved return loss. Further, the percentage bandwidth has significantly improved from 9.58% to 17.79%. The VSWR of the radiating patch array with hybrid HIS-based ground plane is 1.02 at the resonant frequency. Figure 4.19b shows that gain of the planar patch array along the specular direction is 12.42 dB.

However, it can be observed from the contour plot, shown in Figure 4.20, that the maximum gain of 13.2 dB is obtained along θ = 15° and ϕ = 0°. It can be noted that the radiation characteristics of the patch array with conventional ground plane has not degraded when the conventional metallic ground plane is replaced by a hybrid HIS-based ground plane.

The scattering characteristics of the planar patch with hybrid HIS-based ground plane has been further analyzed. As mentioned before, the use of hybrid HIS-based ground plane instead of conventional metallic ground plane reduces the structural RCS of the antenna array. Further analysis is required to assert whether the particular hybrid configuration is able to reduce the radiation mode RCS. Figure 4.21 shows the structural RCS analysis of the planar patch array with hybrid HIS-based ground plane. Figure 4.21a shows the structural RCS pattern at the resonant frequency, i.e., 9.78 GHz. The RCS pattern has a specular lobe of −9.27 dBsm with a side lobe level of −10.8 dB. Figure 4.21b represents the specular structural RCS of the designed patch array with respect to frequency from 8 to 50 GHz. The simulated results are compared with that of planar patch array with conventional ground plane. It can be observed that a wideband (both in-band and out-of-band) reduction in structural RCS is obtained throughout the frequency band considered. In the in-band frequency range, i.e., X-band, a maximum of 2.95 dBsm reduction is obtained (at 11 GHz). The specular RCS value of the planar patch array with hybrid HIS-based ground plane has been reduced maximum by 18.02 dBsm when compared to that of planar patch array with conventional ground plane at 32 GHz. The radiation

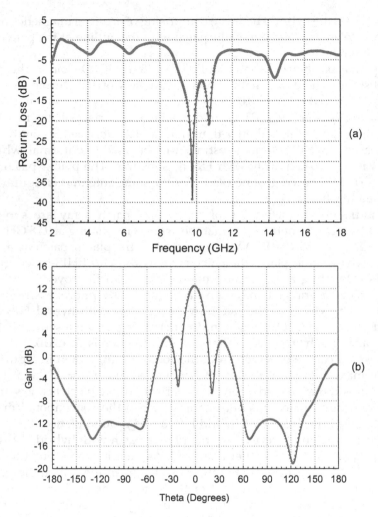

FIGURE 4.19
Radiation characteristics of planar patch array with hybrid HIS-based ground plane: (a) Return loss and (b) gain.

mode RCS of the patch array ($Z_o = 50\ \Omega$; $Z_L = 75\ \Omega$) at the resonant frequency of 9.78 GHz is shown in Figure 4.22. It can be noted that the array RCS has a maximum value of 10.48 dB with a slight offset at $\theta = 1.8°$.

4.3.2 Conformal Patch Array

The planar patch array with hybrid HIS-based ground plane discussed in the previous section has been conformed to a curved surface for radiation and scattering analysis. The curved platform considered is a cylindrical surface with two radii of curvature, i.e., $R = 60$ and $R = 90$ mm as shown in Figure 4.23.

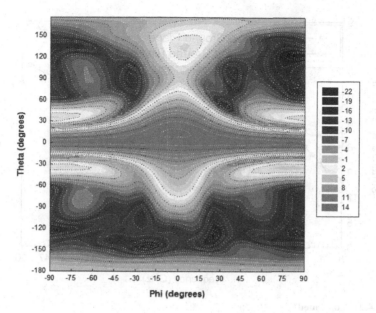

FIGURE 4.20
Gain with respect to theta and phi of a 4-element planar patch array with hybrid HIS-based ground plane.

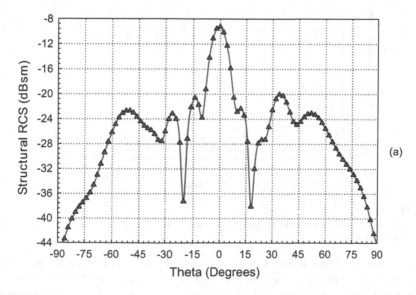

FIGURE 4.21
Scattering characteristics of planar patch array with modified ground plane: (a) Structural RCS pattern. (Continued)

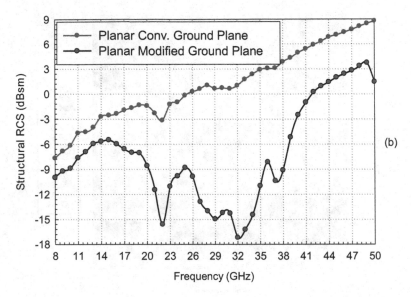

FIGURE 4.21 (Continued)
Scattering characteristics of planar patch array with modified ground plane: (b) specular structural RCS with respect to frequency.

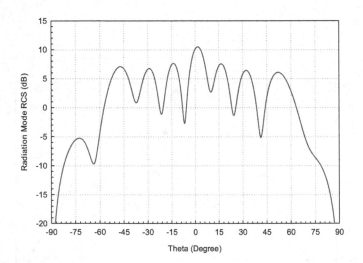

FIGURE 4.22
Radiation mode RCS of 4-element planar patch array with hybrid ground plane; $Z_o = 50\ \Omega$, $Z_L = 75\ \Omega$ at $f = 9.78$ GHz.

The radiation characteristics of the conformal patch array with radius of curvature 60 and 90 mm are compared in Figure 4.24. It can be observed that both the patch arrays resonate at 9.7 GHz. It can be noted from Figure 4.24a that the conformal patch array ($R = 60$ mm) has a return loss of −24 dB with a percentage bandwidth of 19.49 while for conformal patch array with radius of curvature

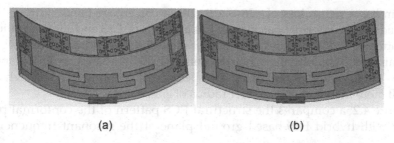

FIGURE 4.23

Bottom view of conformal patch array with hybrid ground plane: (a) $R = 60$ mm and (b) $R = 90$ mm.

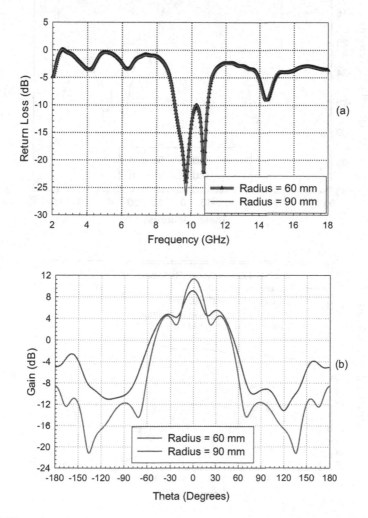

FIGURE 4.24

Radiation characteristics of conformal patch array with hybrid HIS-based ground plane: (a) Return loss and (b) gain.

90 mm has a return loss of −26.47 dB with a percentage bandwidth of 19.48. The radiating conformal patch array has a VSWR of 1.13 and 1.09 for radius of curvature 60 and 90 mm, respectively. It can be observed from Figure 4.24b that the gain of the conformal patch array with hybrid HIS-based ground plane for radius of curvature 90 mm along the specular direction is 10.83 dB, which is 1.7 dB more than that obtained for radius of curvature of 60 mm.

Figure 4.25a compares the structural RCS pattern of the conformal patch array with hybrid HIS-based ground plane at the resonant frequency for

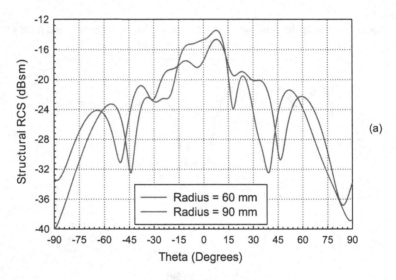

(a)

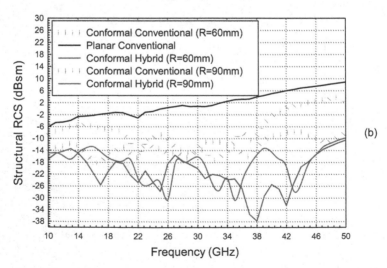

(b)

FIGURE 4.25
Scattering performance of conformal patch array with hybrid HIS-based ground plane: (a) Structural RCS pattern at 9.7 GHz and (b) specular structural RCS with respect to frequency.

radius of curvature 60 and 90 mm. It is apparent that the RCS pattern has the maximum lobe along $\theta = 8°$ for both radius of curvature, i.e., 60 and 90 mm with a value of −14.6 and −13.45 dBsm, respectively. It can be observed that as the radius of curvature decreases, RCS reduction can be further achieved. The specular structural RCS with respect to frequency is shown in Figure 4.25b.

For $R = 60$ mm, a significant reduction in RCS has been obtained in case of conformal patch array with hybrid HIS-based ground plane. A minimum RCS reduction of 9.5 dBsm is observed at 13 GHz while a maximum RCS reduction of 41.7 dBsm is observed at 38 GHz. For $R = 90$ mm, a minimum reduction of 6.37 dBsm at 8 GHz and maximum reduction of 34.48 dBsm at 44 GHz is achieved. It can be observed that the specular structural RCS of conformal patch array with hybrid HIS-based ground plane is less than the planar and conformal patch array with conventional ground plane. The obtained results can be attributed to the curved platform on which the patch array has been mounted.

The radiation mode RCS of the conformal patch array with hybrid HIS-based ground plane has been analyzed at the resonant frequency (9.78 GHz). It can be noted from Figure 4.26 that the antenna RCS has a peak in the specular direction with 10.43 and 11.1 dB for 60 and 90 mm radius of curvature, respectively. The obtained radiation mode RCS in the specular direction is comparable to the respective gain of the conformal patch array.

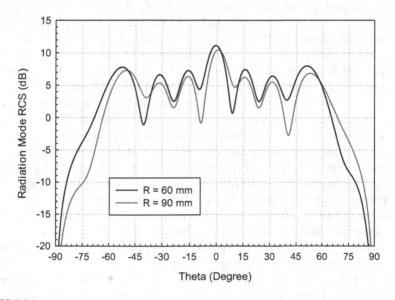

FIGURE 4.26
Radiation mode RCS of 4-element conformal patch array with hybrid ground plane; $Z_o = 50 \; \Omega$, $Z_L = 75 \; \Omega$ at $f = 9.78$ GHz.

4.4 Summary

The contribution of radar cross section by a planar patch array with conventional ground plane increases linearly with the increase in the frequency. Instead, conformal patch arrays are known to have reduced back scattering when compared to planar configurations. The curved platform attributes toward the RCS reduction. For planar patch array configuration, the surface normal of each patch element points toward the same direction (specular) while for a conformal patch array, the surface normal points toward different directions. Hence, the coherent sum of all the scattering contributed by the planar configuration is more than the conformal patch array configuration. This chapter presented conformal patch array with conventional and hybrid HIS-based ground plane for two different radius of curvature, i.e., (i) 60 mm and (ii) 90 mm. The radiation and scattering characteristics of both the conformal patch array are analyzed and compared to the planar patch array with conventional metallic ground plane. It is shown that apart from the slight shift in the resonant frequency, as the radius of curvature of the curved platform increases, there is a reduction in the gain of the conformal patch array. The scattering performance has been analyzed in terms of its structural RCS. A wideband RCS reduction is obtained for both the conformal configurations. However, the extent of RCS reduction increases with increase in the radius of curvature of curved platform. The use of hybrid HIS layer instead of conventional metallic ground plane also contributes to RCS reduction. The performance of such patch array with hybrid HIS based ground plane based on planar and curved platform has been analyzed. In comparison to conformal and planar patch array with conventional metallic ground plane, the structural RCS is reduced over wide band frequencies for conformal patch array with hybrid HIS-based ground plane. Maximum reduction of 41.7 and 34.48 dBsm is attained at 38 and 44 GHz for radius of curvature 60 and 90 mm, respectively. The radiation mode RCS of conformal patch array with the conventional metallic and hybrid HIS ground plane has been computed at the resonant frequency. It has been observed that for conformal patch array with conventional metallic ground plane, the extent of RCS reduction is slightly less for radius of curvature of 90 mm when compared to 60 mm in the X-band. Further, the results show that specular radiation mode RCS is almost equivalent to the gain of the patch array along the specular direction. However, an increase in the side-lobe levels is observed when compared to corresponding planar patch array designs. The hybrid HIS-based ground plane is used to provide out-of-band RCS reduction. However, reducing radiation mode RCS in the in-band frequency range will lead to reduction in the gain of the conformal patch array, which is undesirable. Thus, optimum values of characteristic and load impedance can control the radiation mode RCS in the in-band and out-of-band frequency range. Further, there is a trade-off between the gain and extent of RCS reduction as the radius of curvature of the curved platform increases.

5

Proximity Coupled Patch Array with HIS Layer

Low profile microstrip patch arrays are often preferred in stealth technologies due to their conformal property, ease of fabrication and low weight. One of the disadvantages of patch arrays with microstrip feed is the narrow bandwidth, normally limited to 2%–5%. Thicker substrates with lower dielectric constant can be used to improve the bandwidth. However, as the thickness is made larger, the surface waves and unwanted radiation from the feed network may increase. Further, a corporate feed network in patch arrays consumes a comparatively larger area, when compared to series feed network.

Aperture coupled feeding technique can be introduced to overcome the bandwidth limitation (Balanis, 2005). However, the aperture coupled patch array has a disadvantage of producing backward radiation due to the feed network at the lower side of the patch array. This will lead to reduction in the gain of array. Since the additional layer further complicates the total array structure, the proximity coupled feeding technique can be adopted. This feeding technique provides the maximum impedance bandwidth and minimum spurious radiation (Balanis, 2005).

Numerous techniques have been proposed to reduce the structural RCS of microstrip patch arrays. The use of high impedance surfaces (HIS) has proved to reduce the structural RCS of microstrip patch arrays over wide frequency band as seen in previous chapters. Apart from providing higher impedance bandwidth, aperture coupled feed helps also in reducing structural RCS of a microstrip patch array (Zheng et al., 2008). This is due to the fact that feed network is placed below a dielectric substrate layer; therefore, the contribution of feedlines toward RCS will be minimized. It has been reported that aperture coupled patch antenna with hybrid frequency selective surfaces (FSS)-based ground plane can provide wide band RCS reduction (Cong et al., 2017). A chessboard configuration of two types of split ring resonators and square patches has been used as the hybrid FSS.

In the present work, the proximity coupled feeding technique has been employed to feed microstrip patch array. This has multiple advantages including higher impedance bandwidth, lower backward radiation, and lesser contribution to RCS. Appropriate overlapping of feedline and patch is required for enhancing fundamental resonance. However, the multiple resonances that will be generated from proximity coupled arrays may contribute to higher RCS at corresponding frequencies. Incorporating slots in the ground plane helps in changing the path of surface current, thereby reducing the RCS peaks due to resonant modes. Wu et al. (2008) have proposed a technique to reduce the RCS of patch antenna by cutting slots of varying width in the ground plane. Along with cutting slots in ground plane, loading with chip resistor along the patch has been found to provide low RCS patch antenna (Zheng et al., 2008). A compact resonator consisting of metallic layer with vias connecting to ground plane can be used along the feedline of the patch antenna to suppress the harmonic radiations (Sanchez et al., 2009). The harmonic radiations can also be reduced by varying the dimension and position of the feed (Fistum, 2017).

The RCS reduction feature of HIS and proximity coupled feed network is combined to design a patch array with reduced RCS. This chapter deals with the EM design and analysis of proximity coupled patch array with conventional ground plane, slotted ground plane and hybrid HIS-based slotted ground plane. The radiation and scattering characteristics (both structural and radiation mode RCS) of the patch array in all the mentioned configurations are analyzed.

Proximity coupled feed is a non-contact feeding technique employed in a patch antenna/array. A proximity coupled microstrip patch array is basically a three-layered design with two layers of substrate.

Figure 5.1 shows the schematic of a proximity coupled patch antenna. The radiating patch elements are placed on the top of the upper substrate.

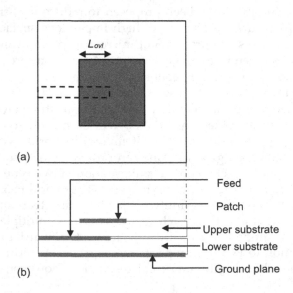

FIGURE 5.1
A proximity coupled patch antenna – Schematic. (a) Top view and (b) side view.

The patch elements are electromagnetically fed through a feed network designed between the two substrates. The coupling between feed and patch is capacitive in nature. The bottom side of the lower substrate has the metallic ground plane. The dielectric constant of both the substrates can be either the same or different, depending on the requirements. Generally, dielectric material with lower permittivity is preferred for upper substrate to enhance radiation into free space, and better efficiency. The bottom substrate is designed with thin dielectric with higher permittivity, so that spurious radiation from feed can be minimized (Balanis, 2005). When compared to coplanar feed, non-contact feed does not have patch-to-feedline discontinuities, and hence it facilitates minimizing of resulting unwanted radiation. The fundamental resonant frequency depends on the extent of overlapping of feedline with the patch. In order to obtain an optimum return loss, the ratio of feed-to-patch overlap (Lovl) to patch length has to be greater than 0.55 (Sanchez et al., 2009).

The proximity coupled feed provides maximum bandwidth of nearly 13%, when compared to coplanar and aperture-coupled feeding techniques (Balanis, 2005). The bandwidth can be improved by terminating the open end of the feedline with a stub. The radiation performance of a proximity coupled patch array can be improved by providing slots in the ground plane. The size and position of the slot affects the performance of the array. The length of the slot is generally taken as $\lambda/2$, while the width to length ratio of the slot is maintained at 1:10. Slots with length comparable to $\lambda/2$ are called resonant slots, while smaller slots are called non-resonant. For maximum coupling, slots are to be placed beneath the center of the patch (Sanchez et al., 2009).

The deployment of slots in the ground plane also helps in reducing RCS of the patch array due to harmonic resonant modes. Incorporating slots causes change in the path of surface current, thereby suppressing the harmonic resonant modes (Wu et al., 2008). Therefore, the contribution of harmonic resonances to the RCS of patch array can be reduced. Vertical slots cut off horizontally directed current, whereas horizontal slots cut off vertically directed current. Therefore, modes having horizontally directed current alone will be suppressed by cutting vertical slots, and vice versa. In this way, the orientation of slots affects the array characteristics. In spite of the advantages mentioned above, the non-contact feeding techniques have the disadvantage of difficulty in multilayer fabrication. There might be chances of misalignment of feed with respect to patch, or creation of gaps between substrate layers.

5.1 Conventional Proximity Coupled Patch Array

Figure 5.2a–d shows the structure of a 4-element corporate-fed proximity coupled patch array with conventional ground plane, designed in X-band. The top layer is comprised of four metallic patch elements on a dielectric

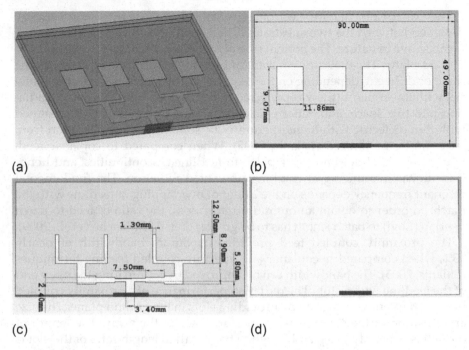

FIGURE 5.2
Proximity coupled patch array with **conventional ground plane**. (a) Perspective view, (b) top layer, (c) middle layer, and (d) bottom layer.

substrate (Figure 5.2b). The patch dimensions are taken as 9.07 × 11.86 mm for 10 GHz resonant frequency. The substrate has dimensions 49 × 90 mm. The patch elements are fed through electromagnetically coupled corporate fed network designed as the middle layer. The feed dimensions are specified in Figure 5.2c.

The overlapping distance of feedline from the edge of the patch is 6 mm. The feed network is sandwiched between two dielectric layers (Figure 5.2c). Both the substrate layers are made of dielectric, RT Duroid (ε_r = 2.2, tan δ = 0.0009) with same thickness (1.58 mm). These layers are backed by a metallic ground plane (Figure 5.2d). Copper (σ = 5.8 × 10^7 S/m) with thickness 0.018 mm is chosen for all metallic parts. The extent of overlapping has significant impact on the fundamental resonant frequency. Figure 5.3 shows the variation of return loss of the proximity coupled patch array with respect to the ratio L_{ovl}/L_p. Here, L_{ovl} represents the length of feed from edge of the patch to the open end of the feed, and L_p is the length of patch element. It can be observed that the proximity coupled patch array is resonant for L_{ovl}/L_p greater than 0.68.

While analyzing the radiation characteristics of the patch array, it can be observed that the array resonates at 10.04 GHz with a return loss of –20.76 dB,

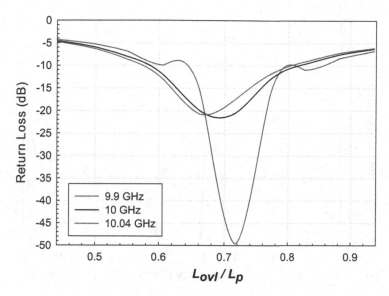

FIGURE 5.3
Return loss of a proximity coupled 4-element linear patch array with respect to L_{ovl}/L_p overlap distance.

as shown in Figure 5.4a. A bandwidth of 6% is obtained for the patch array. The VSWR at the resonant frequency is 1.2. It can be noted from Figure 5.4b that the gain of patch array at $\theta = 0°$ is 7.58 dB, which is relatively small for a 4-element patch array. The side-lobe level is –2.8 dB, and HPBW is 24.3°.

In order to compute the radiation mode RCS of patch array shown in Figure 5.2, the impedance and directivity of the patch elements are extracted by dividing the proximity coupled patch array into segments, each comprised of single patch element and feedline (Table 5.1). These values are used to compute the effective aperture of the patch elements, and then incorporated in the formulation for radiation mode RCS computation.

Figure 5.5a shows the radiation mode RCS with respect to elevation angle computed at 10.04 GHz. The values of both characteristic impedance and load impedance are taken as 50 Ω. It can be noted that the specular RCS at the resonant frequency is 9.45 dB, which is higher than the array gain (i.e., 7.58 dB) as required. The variation of specular radiation mode RCS with respect to frequency in X-band is also analyzed (Figure 5.5b). It is apparent that the specular RCS value increases gradually from 8 to 9.5 GHz, then decreases till 11.5 GHz, and thereafter remains nearly constant. Further, it can be noted that maximum RCS is observed near the resonant frequency. This is in line with expected results. It is to be noted that although the array RCS is decreasing at the upper range of X-band, the radiation performance of the array may be degraded as well.

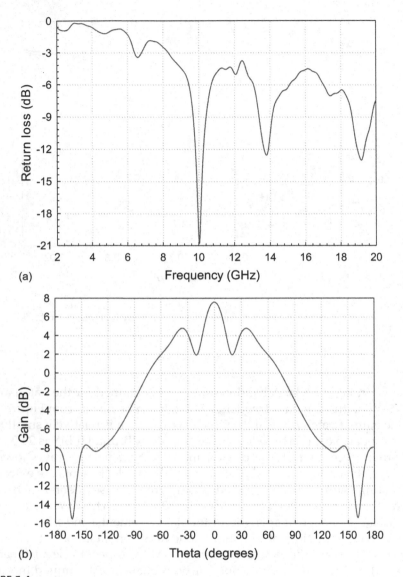

FIGURE 5.4

Radiation characteristics of a 4-element proximity coupled patch array with **conventional ground plane**. (a) Return loss and (b) gain.

The results show that the radiation performance has degraded as compared to two-layered conventional microstrip line fed patch array. This may be due to ineffective coupling of feed to the patch surface. Coupling to the radiating patch elements can be improved by cutting slots in the ground plane. This may help in improving the radiation performance of the patch array.

TABLE 5.1

Antenna Impedance and Directivity of Segments of a 4-Element Proximity Coupled
Patch Array with Conventional Ground Plane

Antenna Segment	Antenna Input Impedance	Directivity
Segment 1	$10.85 + j\,29.36\,\Omega$	4.59
Segment 2	$11.63 + j\,28.85\,\Omega$	4.39
Segment 3	$11.64 + j\,28.84\,\Omega$	4.39
Segment 4	$10.86 + j\,29.35\,\Omega$	4.59

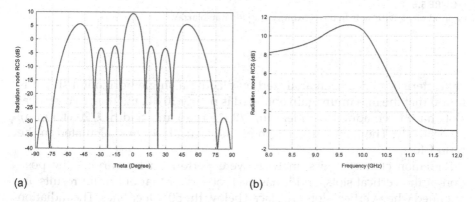

(a) (b)

FIGURE 5.5

Radiation mode RCS of a 4-element proximity coupled patch array with conventional ground
plane; $Z_o = 50\,\Omega$, $Z_L = 50\,\Omega$. (a) RCS pattern at 10.04 GHz and (b) specular RCS with respect to
frequency.

5.2 Gain Enhancement Using Slots

The proximity coupled patch array described in Section 5.1 has been modified by incorporating slots in the ground plane toward the performance
enhancement. Figure 5.6 shows the bottom layer of the 4-element patch array
with the inclusion of slots.

The dimensions of the patch elements and the feed network are same
as those discussed in the previous section. The material of the substrate
and the patch elements are also same as considered before. A horizontal
slot of dimension 59.2 × 1.5 mm has been cut in the ground plane directly
beneath the patch array. The position of the slot is made below the center
of patch length, so as to facilitate maximum coupling. When the radiation

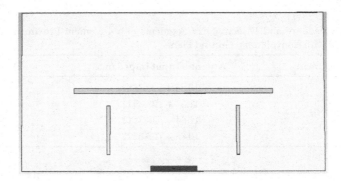

FIGURE 5.6
Ground plane with slots in the proximity coupled patch array.

characteristics have been analyzed by placing a single horizontal slot, it is found that the maximum gain obtained is nearly 10 dB. Further, two vertical slots have been added with length, $\lambda/2$ (15 mm) and width, $\lambda/20$ (1.5 mm), respectively. The width and length of the vertical slot are calculated in the ratio 1:10.

Radiation characteristics analyses were performed by varying the position of the vertical slots, and it has been observed that optimum results are obtained when vertical slots are placed below the 50 Ω feedlines. The radiation characteristics of the proximity coupled patch array are shown in Figure 5.7. It can be observed from Figure 5.7a that the array resonates at 9.76 GHz with a return loss of −33.33 dB. It can further be noted that the bandwidth has improved significantly to 12.94% as compared to the bandwidth of patch array discussed in Section 5.1 (6% bandwidth). The VSWR is obtained as 1.04 (Figure 5.7b). Figure 5.7c shows that the gain of the array is 11.2 dB, which is 3.62 dB higher than the gain of proximity coupled patch array without slots.

This clearly shows the effect of adding slots in increasing the coupling of input power to the patch surface. The array has a side lobe level of −6.7 dB, and HPBW of 20.8°. The variation of gain with respect to theta and phi of the proximity coupled patch array with slotted ground plane is shown in Figure 5.8. The scattering characteristics of the patch array have been analyzed in terms of structural and RCS. The simulated results of the structural RCS are illustrated in Figure 5.9a and b. The monostatic RCS at the resonant frequency (9.76 GHz) for $-90° \leq \theta \leq 90°$ is shown in Figure 5.9a. It can be observed that the specular RCS value is −7.46 dBsm. Figure 5.9b depicts the variation of monostatic specular RCS with respect to frequency. The results are compared with that of a conventional corporate fed patch array of the same specifications. It is evident that reduction in structural RCS is achieved in the frequency range of 10–15 GHz, except near 12.5 GHz. This shows that proximity coupled feeding technique is capable of providing RCS reduction. Maximum

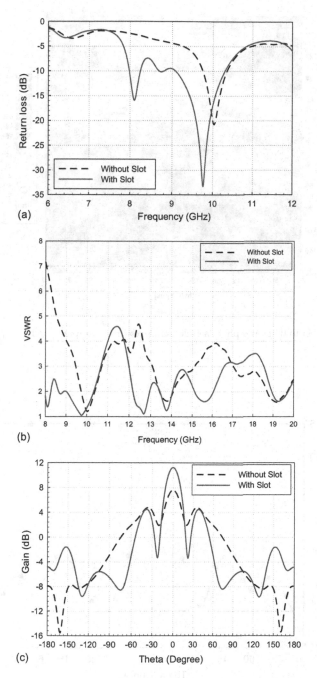

FIGURE 5.7
Radiation characteristics of a 4-element proximity coupled patch array with slotted ground plane. (a) Return loss, (b) VSWR, and (c) gain.

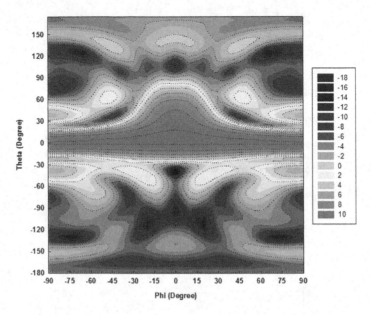

FIGURE 5.8
Gain with respect to theta and phi of a 4-element **proximity coupled** patch array with **slotted** ground plane.

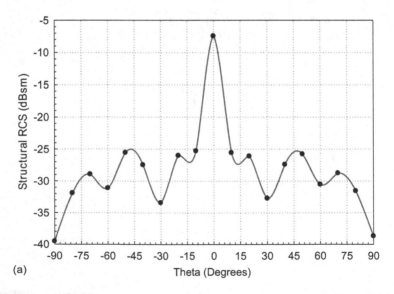

(a)

FIGURE 5.9
Structural RCS of a 4-element proximity coupled patch array with **slotted ground plane**.
(a) Monostatic RCS pattern at 9.76 GHz.   (*Continued*)

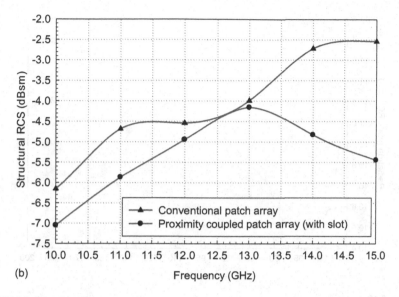

FIGURE 5.9 (Continued)
Structural RCS of a 4-element proximity coupled patch array with **slotted ground plane**.
(b) Specular monostatic RCS with respect to frequency.

reduction of 3 dBsm is obtained at 15 GHz. In order to compute the radiation
mode RCS, the impedance and directivity of the patch elements are extracted
by dividing the patch array structure into four segments. The extracted radiation parameters of these segments are listed in Table 5.2. The difference in
impedance and directivity values between the patch elements can be attributed to the non-uniform nature of the slotted ground plane.

The radiation mode RCS of the proximity coupled patch array with slotted ground plane has been computed for different values of load impedance
(Z_L), while keeping characteristic impedance (Z_o) at 50 Ω (Figure 5.10). It is
evident that the specular RCS goes below the array gain (11.2 dB) for Z_L values between 5 and 130 Ω, which is not desirable. Therefore, Z_L values greater
than 130 Ω have been considered. Figure 5.11a depicts the radiation mode

TABLE 5.2
Antenna Impedance and Directivity of Segments of a 4-Element Proximity Coupled
Patch Array with Slotted Ground Plane

Antenna Segment	Antenna Input Impedance	Directivity
Segment 1	$16.76 + j\,7.68\ \Omega$	3.73
Segment 2	$16.09 + j\,13.98\ \Omega$	1.31
Segment 3	$16.09 + j\,13.98\ \Omega$	1.31
Segment 4	$16.76 + j\,7.68\ \Omega$	3.73

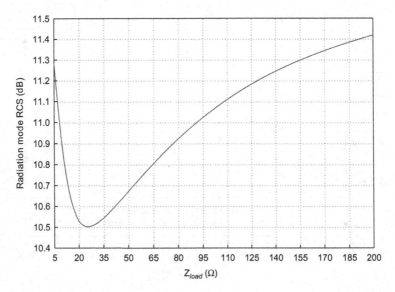

FIGURE 5.10
Variation of specular radiation mode RCS with respect to load impedance (Z_L) of a 4-element proximity coupled patch array with **slotted ground plane.**

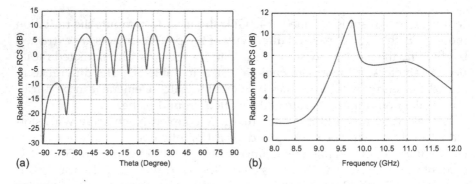

FIGURE 5.11
Radiation mode RCS of a 4-element proximity coupled patch array with **slotted ground plane;** $Z_o = 50\,\Omega, Z_L = 150\,\Omega$. (a) RCS pattern at 9.76 GHz and (b) specular RCS with respect to frequency.

RCS pattern of the patch array with respect to elevation angle at the reso-nant frequency (i.e., 9.79 GHz). The values of Z_o and Z_L are taken as 50 and 150 Ω, respectively. The specular RCS at the resonant frequency is obtained as 11.28 dB. Figure 5.11b shows the specular radiation mode RCS of the patch array with respect to frequency for Z_o and Z_L values of 50 and 150 Ω, respec-tively. It can be observed that the RCS increases exponentially from 8 GHz to reach a maximum value near the resonant frequency, beyond which the RCS reduces.

5.3 RCS Reduction Using HIS Layer

The patch array design to be discussed in this section is a modification of the design already described in previous section. The ground plane of the proximity coupled patch array (Figure 5.6) is replaced with a reduced slotted ground plane. Figure 5.12a and b shows the modified structure of the proximity coupled patch array.

The HIS layer consists of alternately oriented 2 × 2 array of Jerusalem cross (JC) elements and a metallic square patch (Figure 5.12a). The dimensions of these elements are taken from the design of two-layered microstrip patch array with hybrid HIS-based ground plane as discussed in previous chapters. A rectangular metallic patch is designed on ground plane region that comes beneath the patch array. A horizontal slot of dimension, 59.2 × 1.5 mm and two vertical slots, each of dimension, 15 × 1.5 mm are cut on this rectangular patch on the ground plane. The ground plane is reduced in a way such

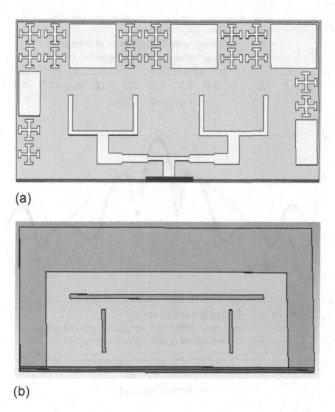

(a)

(b)

FIGURE 5.12
Proximity coupled patch array with **hybrid HIS layer and reduced slotted ground plane**. (a) Middle layer and (b) bottom layer.

that the radiation characteristics of the proximity coupled patch array with conventional slotted ground plane (Figure 5.6) are not disturbed.

The radiation characteristics of the proximity coupled patch array with hybrid HIS-based slotted ground plane are analyzed. A comparison between these results with the radiation characteristics of proximity coupled patch array with and without slotted ground plane is illustrated in Figure 5.13.

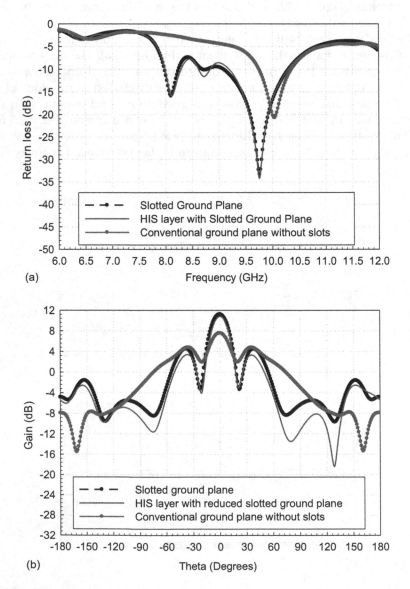

(a)

(b)

FIGURE 5.13
Radiation characteristics of a 4-element proximity coupled patch array with **hybrid HIS layer and reduced slotted ground plane.** (a) Return loss and (b) gain.

It is evident that the radiation performance of patch array with slotted ground plane is not much affected by incorporating the HIS chessboard configuration in the middle layer. It can be observed from Figure 5.13a that the patch array with hybrid HIS-based slotted ground plane resonates at the same frequency (i.e., 9.76 GHz) as that of conventional slotted ground plane. The return loss at the resonant frequency is obtained to be −31.32 GHz, and has a percentage bandwidth of 11.7.

The designed patch array shows a gain of 10.8 dB at $\theta = 0°$, which is only 0.4 dB lower than the corresponding gain of patch array with slotted ground plane (Figure 5.13b). The side lobe level has reduced to −7.4 dB, and the HPBW is 19.4°, which is slightly lower than the HPBW of patch array with slotted ground plane.

The scattering characteristics of the patch array have been analyzed in terms of monostatic structural RCS. The simulated results of the structural RCS are illustrated in Figure 5.14. The monostatic RCS pattern has been simulated at their respective resonant frequencies for −90° ≤ θ ≤ +90°. It can be observed from Figure 5.14a that the structural RCS values of the 4-element proximity coupled patch array with slotted ground plane and HIS layer with reduced slotted ground plane are −7.46 and −7.85 dBsm, respectively. Further, the specular structural RCS of all the proximity coupled patch array is simulated with respect to frequency (Figure 5.14b) and compared to that of conventional metallic ground plane (reference antenna). It can be observed

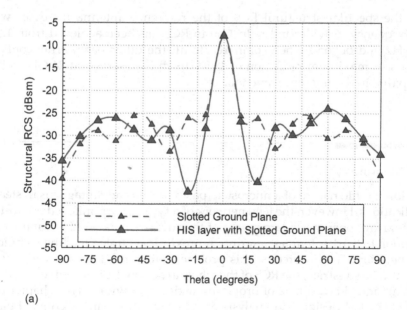

(a)

FIGURE 5.14
Structural RCS of a 4-element proximity coupled patch array with **hybrid HIS layer and reduced slotted ground plane**. (a) Monostatic RCS pattern at 9.76 GHz. *(Continued)*

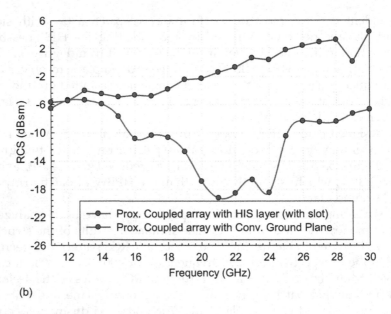

(b)

FIGURE 5.14 (Continued)
Structural RCS of a 4-element proximity coupled patch array with **hybrid HIS layer and reduced slotted ground plane**. (b) Specular monostatic RCS with respect to frequency.

that the specular structural RCS of the reference antenna increases with the frequency. A wideband reduction in RCS has been achieved from 12 to 30 GHz and beyond when compared to all the other patch array configurations. A maximum reduction of 18.83 dBsm is observed at 24 GHz when compared to the reference antenna.

5.4 Summary

The low profile nature of a microstrip patch array is advantageous in stealth applications. However, the feed network of a typical corporate fed microstrip patch array consumes a larger area of the surface. This will have more contribution toward antenna scattering. A proximity coupled feeding technique can be adopted to overcome this problem. Since the feed network comes beneath the substrate, the RCS of the whole array can be reduced to an extent, with an added advantage of providing wider bandwidth. This chapter presented the EM design and analysis of a 4-element proximity coupled patch array with three different ground plane configurations such as (i) conventional ground plane, (ii) slotted conventional ground plane, and (iii) hybrid HIS-layer with reduced slotted ground plane. The radiation and scattering characteristics of all the configurations of patch array have been analyzed. The results show that cutting slots in the ground plane of a proximity coupled patch array has aided in gain enhancement. The patch array with slotted ground plane has shown a· gain of 11.2 dB, which is 3.62 dB higher than that of conventional proximity coupled patch array. Moreover, the percentage bandwidth has doubled (6%–12.94%) when slots were added to the ground plane of conventional proximity coupled patch array. Further, it has been observed that the radiation performance of the proximity coupled patch array with reduced slotted ground plane has not degraded by the inclusion of HIS chessboard configuration in the middle layer. The structural RCS of proximity coupled patch array with slotted ground plane is found to be lower than a conventional patch array over wide band. This demonstrates the ability of proximity coupled patch array to reduce RCS. The structural RCS is further reduced by using a proximity coupled patch array with hybrid HIS layer based slotted ground plane. The radiation mode RCS of proximity coupled patch array with slotted ground plane is reduced in the lower range of X-band, as compared to similar patch array without slots.

6

Conclusion

Antenna scattering has been a complex problem due to the fact that reducing RCS of an antenna, especially when it is in radiating mode, would have adverse effect on its radiation performance. There are practical situations in aerospace/naval applications when the antenna/array RCS dominates over RCS of the platform over which it is mounted. Since radiation mode RCS is almost two orders of magnitude higher than its structural RCS, it is very important to control antenna scattering while it is radiating. The antenna mode RCS can be minimized in matched condition (in between radiating structure and its feed). However, it would lead to degradation in the antenna gain, especially at resonant frequency. The designs of different configurations of patch array discussed throughout the book demonstrate similar concepts.

In this book, HIS-based modified ground plane in the antenna structure has been exploited toward achieving antenna RCS reduction (both structural RCS and radiation mode RCS). It is the type of the AMC element used in the design that controls in-band and out-of-band RCS reduction. Thus, the array designs presented in the book can be further exploited by incorporation of active elements in the feed network or implementing efficient adaptive algorithms in array processing to achieve reduction in array RCS even when it is in its radiation mode. This is referred to as active RCS reduction. This technique would place nulls in the RCS pattern toward the probing directions. Thus, the phased array becomes invariably invisible to the probing radar.

References

Balanis, C.A., *Antenna Theory, Analysis and Design*. Hoboken, NJ: John Wiley & Sons, ISBN: 0-471-66782-X, 1117 p., 2005.

Cong, L., X. Cao, and T. Song, "Ultra-wideband RCS reduction and gain enhancement of aperture-coupled antenna based on hybrid-FSS," *Radio Engineering*, vol. 26, no. 4, pp. 1041–1047, December 2017.

Costa, F., A. Monochio, and G. Manara, "An overview of equivalent circuit modeling techniques of frequency selective surfaces and metasurfaces," *ACES Journal*, vol. 29, no. 12, pp. 960–976, December 2014.

Fistum, D., "Efficient proximity coupled feed rectangular microstrip patch antenna with reduced harmonic radiation," *Indonesian Journal of Electrical Engineering and Computer Science*, vol. 7, no. 2, pp. 500–506, August 2017.

Hansen, R.C., "Relationship between antennas as scatterers and as radiators," *Proceedings of the IEEE*, vol. 77, no. 5, pp. 659–662, May 1989.

Iriarte, J.C., M. Paquay, I. Ederra, R. Gonzalo, and P.D. Maagt, "RCS reduction in a chessboard-like structure using AMC cells," *Proceedings of European Conference on Antennas and Propagation*, pp. 1–4, November 11–13, 2007.

Iriarte, J.C., A.T. Pereda, J.L.M.D. Falcón, I. Ederra, R. Gonzalo, and P.D. Maagt, "Broadband radar cross-section reduction using AMC technology," *IEEE Transactions on Antennas and Propagation*, vol. 61, no. 12, pp. 6136–6143, December 2013.

Jenn, D.C., *Radar and Laser Cross-Section Engineering*. Washington, DC: AIAA Education Series, ISBN: 1-56347-105-1, 476 p., 1995.

Josefsson, L. and P. Persson, *Conformal Array Antenna Theory and Design*. Hoboken, NJ: John Wiley & Sons, ISBN: 4978-0-471-46584-3, 488 p., 2006.

Li, M., S.Q. Xiao, Y.Y. Bai, and B.Z. Wang, "An ultrathin and broadband radar absorber using resistive FSS," *IEEE Antennas and Wireless Propagation Letters*, vol. 11, pp. 748–751, 2012.

Li, Y., Y. Liu, and S.-X. Gong, "Microstrip antenna using ground-cut slots and miniaturization techniques with low RCS," *Progress in Electromagnetics Research Letters*, vol. 1, pp. 211–220, 2008.

Liu, Y., H. Wang, Y. Jia, and S.-X. Gong, "Broadband radar cross-section reduction for microstrip patch antenna based on hybrid AMC structures," *Progress in Electromagnetics Research C*, vol. 50, pp. 21–28, 2014.

Munk, B.A., *Frequency Selective Surfaces, Theory and Design*. 1st ed., New York: John Willey & Sons, ISBN: 0-471-37047-9, 410 p., 2000.

Narayan, S., B. Sangeetha, and R.M. Jha, *Frequency Selective Surfaces based High Performance Microstrip Antenna*. Singapore: Springer, Springer Briefs in Electrical and Computer Engineering-Computational Electromagnetics, ISBN: 978-981-287-774-1, 45 p., 2016.

Paquay, M., J.C. Iriarte, I. Ederra, R. Gonzalo, and P.D. Maagt, "Thin AMC structure for radar cross-section reduction," *IEEE Transactions on Antennas and Propagation*, vol. 55, no. 12, pp. 3630–3638, December 2007.

Sanchez, L.I., J.L.V. Roy, and E.R Iglesias, "Proximity coupled microstrip patch antenna with reduced harmonic radiation," *IEEE Transactions on Antennas and Propagation*, vol. 57, no. 1, pp. 27–32, January 2009.

Sievenpiper, D., L. Zhang, R.F.J. Broas, N.G. Alexopolous, and E. Yablonovitch, "High-impedance electromagnetic surfaces with a forbidden frequency band," *IEEE Transactions on Microwave Theory and Techniques*, vol. 47, no. 11, pp. 2059–2074, November 1999.

Singh, A. and H. Singh, "Active cancellation of probing signals in dipole array mounted on conformal platforms," *IEEE Indian Conference on Antennas and Propagation* (InCAP 2018), Hyderabad, India, December 16–19, 2018, 2 p., 2018.

Singh, H., R. Chandini, and R.M. Jha, *RCS Estimation of Linear and Planar Dipole Phased Arrays: Approximate Model*. Singapore: Springer, Springer Brief in Electrical and Computer Engineering-Computational Electromagnetics, ISBN: 978-981-287-753-6, 47 p., 2015a.

Singh, H. and R.M. Jha, *Active Radar Cross Section Reduction: Theory and Applications*. Cambridge, UK: Cambridge University Press, ISBN: 978-1-107-092617, 325 p., 2015.

Singh, H., H.L. Sneha, and R.M. Jha, *Radar Cross Section of Dipole Phased Arrays with Parallel Feed Network*. Singapore: Springer, Springer Brief in Electrical and Computer Engineering-Computational Electromagnetics, ISBN: 978-981-287-783-3, 77 p., 2015b.

Wu, T., Y. Li, S.X. Gong, and Y. Liu, "A novel low RCS microstrip antenna using aperture coupled microstrip dipoles," *Journal of Electromagnetic Waves and Applications*, vol. 22, pp. 953–963, ISSN: 1569-3937, 2008.

Zheng, J.H., Y. Liu, and S.X. Gong, "Aperture coupled microstrip antenna with low RCS," *Progress in Electromagnetics Research Letters*, vol. 3, pp. 61–68, 2008.

Index

Note: Page numbers in italic and bold refer to figures and tables, respectively.

Printed in the United States
by Baker & Taylor Publisher Services

Printed in the United States
by Baker & Taylor Publisher Services